About the Author

Joseph B. Dym is a professional engineer licensed to practice mechanical engineering and a member of SPE. His experience spans a period of 55 years and covers all aspects of design engineering, as well as related fields such as tooling and manufacturing. As managing engineer, he was involved not only in the design, but also in product development, prototyping, testing, or verifying design requirements of finished products — with a sizable portion of activity in plastic items.

The author has been Chief Engineer of the Mine Safety Appliances Co.; Manager of the Corporate Manufacturing Engineering Department — an advisory group of engineers — at Westinghouse Electric Corporation; Chief Engineer and Plant Manager of a custom molding shop; and Chief Engineer of Plastics at Rockwell International Corporation, Uniontown, Pennsylvania. Currently retired, he remains active as a consultant in the areas of plastic part design, mold design, and processing plastics.

Jacket design by Steve Baron

Product Design with Plastics
A Practical Manual

JOSEPH B. DYM, P.E.

PRODUCT DESIGN WITH PLASTICS

A PRACTICAL MANUAL

INDUSTRIAL PRESS INC.
200 Madison Avenue
New York, NY 10157

Library of Congress Cataloging in Publication Data

Dym, Joseph B.
    Product design with plastics.

    Includes index.
    1. Plastics.   2. Design, Industrial.   I. Title.
TA455.P5D95   620.1'923   82-3091
ISBN 0-8311-1141-0   AACR2

SECOND PRINTING
PRODUCT DESIGN WITH PLASTICS: A PRACTICAL MANUAL

To My Family

Acknowledgments

I wish to express my sincere appreciation to the following companies for their response to my requests for data and technical information on their materials: American Cyanamid Co.; American Hoechst Corp.; Amoco Chemicals Corp.; Borg Warner Chemicals Inc.; Celanese Plastics and Specialties Co.; Ciba-Geigy Corp.; Du Pont de Nemours & Co.; Eastman Plastics; Fiberite Corp.; General Electric Co.; B. F. Goodrich Co.; Hercules, Inc.; Hooker Chemicals and Plastics Corp.; LNP Corp.; Mitsui Chemicals; Mobay Chemical Co.; Monsanto Co.; Phillips Chemical Co.; Plaskon Products Inc.; PPG Industries; Rogers Corp.; Rohm and Haas Co.; 3M Co.; Union Carbide Corp.; Uniroyal Corp.

Preface

Product Design with Plastics: A Practical Manual was written to show designers various methods of applying commercial grades of plastic materials and also of using the available information on the characteristics of such materials. The information or data developed on the materials are derived from readily available facilities and are relatively inexpensive. Should a need arise for data at conditions different from those at which tests are available, it would not be too difficult or costly to obtain the needed information, since it would merely involve adjusting the existing facilities to the specified requirements. Thus the data supplied by the manufacturers of materials or additional data when needed should be the basis on which product designs are evolved. This approach is backed by personal experience extending over 50 years during which a great variety of designs with plastic materials were resolved.

Design problems with conventional materials of construction are usually solved with the aid of textbooks or handbooks that refer the reader to data sheets where the characteristics of a specific material are listed. However, parts designed with plastics involve some special considerations. Plastic material suppliers provide material data sheets for each grade they produce.

At first glance, there could be a tendency to apply the plastic information in a similar fashion to that of other materials. If such a procedure were to be followed, the result would not only lead to disappointment, but perhaps even to product failure. The reason for the difference in treating the plastic data sheets from those of other materials is the behavior of plastics under load and under varying environmental conditions, which normally are not factors with other materials.

The reaction of plastics under test conditions is explained in Chapters 2 and 3 where the test procedure, the conditions under which the tests are conducted, and interpretation of the results are fully explained. Until this phase of the data sheets is properly understood, it is best not to apply the numbers from the data sheet, for they can only be a source of misinterpreted information. A considerable segment of the data is only usable for comparative evaluation of various grades of material. Even in this case one must be sure that the test procedure and the test conditions were the same. In many cases, product use conditions are very different from data sheet test conditions; therefore, it would be unsafe to attempt interpolating the available information, and the needed data should be obtained under conditions simulating end use. The reader should recognize that knowledge of plastic material tests is a prerequisite to understanding the meaning and value of data for design purposes as presented in the suppliers' sheets.

Another factor to consider in the early stages of design is material selection in relation to cost per cubic inch. Since the material value in a plastic part is roughly one-half of its overall cost, it becomes important to select a candidate material with extraordinary care.

After a material type and grade that will fulfill performance requirements has been decided upon, steps should be taken to ensure that degrading features such as inside sharp corners, nonuniform wall thicknesses, etc., are eliminated from the design details. It should be noted that the features that can degrade properties have a much greater effect on plastics than on other materials.

The design of a part is not the only area where its properties can be degraded; this can also occur in the areas of tooling and processing. Therefore, the designer must be involved in every phase of product creation so that the desired characteristics are reproduced in the finished parts.

Another major consideration in design is the application of the correct numerical values for various properties as indicated in some practical examples in Chapter 7.

In all other respects, normal design procedure can be followed. When all the above-mentioned considerations are kept in mind, one can expect plastic parts to perform to the full satisfaction of the user.

This volume follows the general pattern of a "strength of materials" text and does not cover the role of polymer structure in relation to physical properties and similar aspects, since the designer can work only with materials of certain properties as produced by the manufacturer of the raw product and can do little if anything to modify those properties. Under such circumstances the basic material composition is of no direct concern to the designer. However, Chapter 4, dealing with polymer formation, is included for the reader's general information. Finally, in addition to taking into account all of the relevant elements that will ensure a sound product design, it must be kept in mind that prototype testing to verify performance is an important step in the overall design process.

This volume will be of use not only to designers but also to drafting personnel, tool designers, and anyone involved in various stages of processing plastic materials.

The combination of Index and Glossary should prove convenient to anyone connected with the plastics industry who is looking for the meaning of terms and of tradenames in this field.

Sincere thanks are expressed to C. H. Basdekis of Monsanto Plastics & Resins Co. for his critical review of Chapters 4 and 5, covering plastic materials. His constructive additions and modifications helped to clarify certain sections in these chapters.

Contents

Product Design with Plastics
A Practical Manual

1. Introduction to the Application of Plastics

Plastic materials are predominantly synthetic materials. In the last 30 years plastics have enjoyed a growth that has been unequaled by any other group of materials. This demand continues to increase, and the facilities for meeting the requirements are being expanded continuously.

There have to be good reasons for the phenomenal application of plastics in order to justify the large investments needed to produce the raw materials and to convert them into finished products. Some of the characteristics that recommend these plastic materials to designers and users will be discussed in this chapter.

Overall, it can be stated that plastic products meet the following criteria: their functional performance meets use requirements; they lend themselves to esthetic treatment at comparatively low cost; and , finally, the finished product is cost competitive.

Some of the specific properties that are determining factors when considering the use of plastic materials are listed below:

1. Weight.

Their average weight is roughly one-eighth that of steel. In the automotive industry, where lower weight means more miles per gallon of gasoline, the utilization of plastics is increasing with every model-year. For portable appliances and portable tools lower weight helps to reduce the fatigue factor. Lower weight is beneficial in shipping and handling, and as a safety factor to humans.

2. Thermal and electrical insulation.

The value of heat insulation is fully appreciated in the use of plastic drinking cups and of plastic handles on cooking utensils, electric irons, and other devices where heat can cause discomfort or burning. In electrical devices the plastic material's application is extended to provide not only voltage insulation where needed, but also the housing that would protect the user against accidental grounding. In industry the thermal and electrical uses of plastics are many, and these uses usually combine additional features that prove to be of overall benefit.

3. Corrosion resistance and color.

Protective coatings for plastics are not required owing to the inherent corrosion-resistant character of this material. The eroding effects of rust are well known, and a material that does not deteriorate offers distinct advantages. Colors for esthetic appearance are incorporated in the material compound and become an integral part of the plastic for the life of the product.

4. Part features.

Combining bosses, ribs, and retaining means for assembly are easily attained in plastic parts, resulting in manufacturing economies that are frequently used for cost reduction.

5. High volume production.

Plastic conversion into finished products for large volume needs has proven to be one of the most cost-effective processes. It is a case where technology shines.

6. Transparency.

When transparency is needed in conjunction with toughness, plastic materials are the preferred candidates.

7. Coefficient of friction.

Many plastic materials inherently have a low coefficient of friction. Other plastic materials can incorporate this property by compounding a suitable ingredient into the base material. It is an important feature for moving parts, which provides for self-lubrication.

8. Chemical resistance.

Chemical resistance is another characteristic that is inherent in.plastic materials. The range of this resistance varies among materials.

Materials that have all these favorable properties also have their limitations. Every designer reading this list of outstanding characteristics would ask: How can I apply these materials and take advantage of their benefits? The answer has to be with caution, care, and attention to detail. Specifically this means that one must find out what measures are taken to evaluate the materials, carefully study the test results and test methods that are employed in obtaining these properties and their interpretation for application purposes, and finally determine the fine details of use conditions to establish the suitability of a plastic material for the intended purpose.

The first few chapters are arranged to give the designer a brief description of short- and long-term tests, and their meaning and applicability for design purposes. These chapters are followed by an explanation of how the plastic raw material is formed and, subsequently, by data that are descriptive of each material family, as well as by property information on grades within each family.

This basic information on materials leads to design consideration for plastics and pertinent information related to elements of producing them. Each chapter is self-contained and can be read without following the outlined sequence. An overall concept of designing with plastic can be best attained by a systematic review of the outlined program.

$2.$ Description and Derivation of Short-Term Properties

In the early stages of product design, one visualizes a certain material, makes approximate calculations to see if the contemplated idea is practical, and, if the answer is favorable, proceeds to collect detailed data on a range of materials that may be considered for the new product. When plastics are the candidate materials, it must be recognized from the beginning that the available test data require understanding and proper interpretation before an attempt can be made to apply them to part design. For this reason, an explanation of data sheets is given in order to avoid anticipating part characteristics that may not exist when merely applying data sheet information without knowing how such information was derived.

PLASTICS PROPERTY DATA

The application of appropriate data to product design can mean the difference between the success and failure of an item manufactured from plastics. There are two important sources of information on plastics.

First, the data sheet compiled by a manufacturer of the material and derived from tests conducted in accordance with specifications outlined by the American Society for Testing of Materials (ASTM). (These ASTM tests will be explained in this chapter.)

Second, the description of outstanding characteristics of each polymer (plastic material), along with the listing of typical applications. (This second source will be detailed in the following chapter, where data sheets are integral parts of the material description.)

These two information sources should provide the designer the necessary basis with which he or she can engineer a plastic product that would perform successfully in service.

It is obvious that it is important for the designer to become familiar with all the information that is available for each plastic, especially that which is pertinent to product design. Engineers and designers are knowledgeable of the data derived from metal tests, and there could be a tendency to apply the plastic data sheet information in a manner similar to that used for metals. This could be understood because there is no warning that some of the data supplied by the manufacturer are applicable only when *use* and *test* conditions are nearly the same. The only hint of caution is the citing of the ASTM number alongside each property. However, if suppliers' data were to be applied without a complete analysis of the test data for each property, the result could prove costly and embarrassing. The nature of plastic materials is such that an oversight of even a small detail in its properties or the method by which they were derived could result in use problems.

Once it is recognized that there are certain reservations with some of the properties given on the data sheet, it becomes obvious that it is very important for the designer to have a good understanding of these properties for the interpretation of the ASTM test results in order to make the proper strength calculations and to select a material for a specific item.

Taking into account the necessary knowledge and calculations related to a plastic material, it still appears advisable, or one can say mandatory, that product use tests be extensively performed to verify expected results. Actual material data sheets can be found in the following chapter. Some data sheets may omit certain property data, thereby implying that the specific material is not deemed suitable for an application for which the test information is omitted.

It will be the aim of this chapter to explain the meaning of the data derived from the ASTM tests, especially as they relate to design parameters, and to emphasize the degree of their applicability. The description of test procedure and specimen details will be abbreviated to merely convey the concept of the tests.

The sequence of the test numbers follows the one arranged on the data sheets in Chapter 3. Anyone interested in all the details of the ASTM tests will find them in *Plastics—General Methods of Testing, Nomenclature*. This is one of the many parts of *ASTM Standard* and is available in libraries or directly from the ASTM or the U.S. Government Printing Office.

In this chapter, the headings will, in general, conform to those of the ASTM test, but the property test number will be indicated in parentheses in order to emphasize that it is given for reference purposes only.

A. PHYSICAL PROPERTIES—GENERAL

1. SHRINKAGE FROM MOLD DIMENSIONS OF MOLDED PLASTICS (ASTM-D955)

Specimen and Procedure

The specimen for compression and transfer molding is a test bar $\frac{1}{2}'' \times \frac{1}{2}'' \times 5''$ and a round disc $\frac{1}{2}''$ thick $\times 2''$ or $4''$ in diameter. For injection molding, the test bar is $\frac{1}{2}'' \times 5'' \times \frac{1}{8}''$ thick.

The molding parameters are outlined by the material's manufacturer. They are of little interest to the designer since a product's molding conditions hardly ever approach the molding procedure of this test. Shrinkage measurements are usually made 48 hours after the specimen has been molded. The test results show the shrinkage of the moldings against the dimensions of the mold cavity.

Meaning of Data

The use of correct shrinkage information is very important, not only for having the desired proportions of a product, but also for fitting with other parts and for functional purposes. The shrinkage data are usually shown in a range of two values. The lower figure is intended to apply to thin parts (0.70″ or less), whereas the higher figure would involve thicknesses of ⅛″ or more.

The choice of shrinkage for a selected material and a specific design is the responsibility of the mold designer, molder, and product designer. If experience with the selected grade of material is limited, the design should be submitted to the material supplier for recommendations, and the data coordinated with the interested parties. Where very close tolerances are involved, preparing a prototype of the part may be necessary to establish critical dimensions. If this step is not practical, it may be necessary to test a mold during various stages of cavity fabrication with allowances for correction—in order to determine the exact shrinkage needed.

Considering the factors that can contribute to variations in shrinkage, it will be fully appreciated how significant it is to select the appropriate numbers.

The thermosetting compression-molded parts will have a higher shrinkage

a. when cavity pressure is on the low side;

b. when mold temperature is on the high side;

c. when cures are shorter;

d. when parts are thicker (over 3/16″);

e. when a material is soft flowing (highly plasticized);

f. when a material is preheated at relatively low heat;

g. when a high moisture content is present in the raw material.

Transfer-molded and injection-molded thermosetting resins will have higher shrinkages in comparison with compression-molded parts of the same design and material. Some of those higher shrinkages are due to the imparted directional flow and others are due to a tendency to use small gates that do not permit the application of higher pressures to the cavity.

The shrinkage of injection-molded thermoplastics will be affected as follows:

a. Higher cavity pressures will cause lower shrinkages.

b. Thick parts (⅛″ or more) will shrink more than thin ones.

c. A cool mold [80°F (27°C) or less] will bring about a lower shrinkage, whereas a mold temperature of 120°F (49°C) or more will produce higher shrinkages.

d. A melt temperature of the material at the lower end of the recommended range will produce a lower shrinkage and one at the upper end of the range will produce a higher shrinkage.

e. A longer cycle time, above the required solidification point, will partially conform the part closer to the mold dimensions, thereby bringing about lower shrinkage.

f. Openings in a part (holes and core shapes) will somewhat interfere with shrinkage in the cavity and thus will bring about lower and varying shrinkages than would be the case in a part without openings.

g. Larger feeding gates to the part will cause lower shrinkage permitting higher pressure buildup in the cavity.

h. Materials that are crystalline or semicrystalline have dual shrinkage: higher in the direction of material flow and lower perpendicular to it. In a symmetrical part, when center gated, the shrinkage will average out and be reasonably uniform.

i. Most thermoplastics attain their full shrinkage after 24 hours, but there are some which may take weeks to stabilize their dimensions fully. The manufacturer of the material usually indicates whether there is a delayed shrinkage effect present.

j. Glass-reinforced or otherwise filled thermoplastics have considerably lower shrinkages than the basic polymer.

The data on shrinkage have to be approached with much care if one is to avoid dimensional problems with the plastic product.

2. SPECIFIC GRAVITY AND DENSITY (ASTM-D792)

Specimen and Procedure

This property can be determined on moldings, sheet, or rod. A piece of the specimen is attached to a fine wire, weighed, and submerged in water. While in the water it is weighed again. From the difference in weight, the density can be calculated.

Meaning of Data

Specific gravity and density are frequently used interchangeably; however, there is a very slight difference in their meaning.

Specific gravity is the ratio of the weight of a given volume of material at 73.4°F (23°C) to that of an equal volume of water at the same temperature. It is properly expressed as "specific gravity, 23/23°C."

Density is the weight per unit volume of material at 73.4°F (23°C) and is expressed as "D23C, g per cm^3."

The discrepancy enters because water at 73.4°F (23°C) has a density slightly less than one. To convert specific gravity to density, the following factor can be used:

$$\text{D23C, g per cm}^3 = \text{specific gravity, 23/23°C} \times 0.99756$$

To the designer, the specific gravity is useful in calculating strength−weight and cost−weight ratios, and as a means of identifying a material.

Specific volume is a conversion of specific gravity into cubic inches per pound. Since the volume of material in a part is the first bit of information established after its shape is formulated, the specific volume is a convenient conversion factor for weight:

$$\text{specific volume (in cubic inches per pound)} = \frac{27.7}{\text{specific gravity}}.$$

3. WATER ABSORPTION OF PLASTICS (ASTM-D570)

Specimen and Procedure

For molding materials a disc of 2″ diameter × ⅛″ thick and for sheet material a bar of 3″ × 1″ wide × thickness of sheet are used. The specimens are dried for 24 hours in an oven at 122°F (50°C), cooled in a desiccator, and weighed immediately. The test is usually an immersion of the specimen for 24 hours (longer if desired) at 73.4°F (23°C) in distilled water. Upon removal, the specimen is wiped dry with a cloth and weighed immediately.

Meaning of Data

The data should indicate the temperature and time of immersion and the percentage of weight gain. The same applies to data at the saturation point of 73.4°F (23°C), and, if the material is usable at 212°F (100°C), also to saturation at this temperature.

Moisture or water absorption is an important design property. It is particularly significant for a part that is used in conjunction with other materials which call for fits and clearances along with other close tolerance dimensions.

The moisture content of a plastic affects electrical insulation resistance, dielectric losses, mechanical properties, dimensions, and appearances. The effect on the properties due to moisture content depends largely on the type of exposure (by immersion in water or by exposure to high humidity), the shape of the part, and the inherent properties of the plastic material. The ultimate proof for tolerance of moisture in a part has to be a product test under extreme conditions of usage in which critical dimensions and needed properties are verified.

4. HAZE AND LUMINOUS TRANSMITTANCE OF TRANSPARENT PLASTICS (ASTM-D1003)

Specimen and Procedure

Transparent film and sheet samples, or molded samples which have parallel plane surfaces, are used. A disc of 2″ diameter is recommended. No conditioning is required.

Procedure A is followed when a hazemeter is used in the determinations. Procedure B is followed when a recording spectrophotometer is used (known as a GE Spectrophotometer).

Meaning of Data

In this test, *haze* of a specimen is defined as the percentage of transmitted light that, in passing through the specimen, deviates more than 2.5° from the incident beam by forward scattering.

Luminous transmittance is defined as the ratio of transmitted to incident light.

These qualities are considered in most applications for transparent plastics, forming a basis for directly comparing the transparency of various grades and types of plastic.

The data are of value when a material is considered for optical purposes. Many transparent plastics do not have water clarity, and, for this reason, the data should indicate whether the material was natural or tinted when tested.

5. LUMINOUS REFLECTANCE, TRANSMITTANCE, AND COLOR (ASTM-D791)

Specimen and Procedure

Opaque specimens should have at least one plane surface. Translucent and transparent specimens must have two surfaces which are plane and parallel. The specimen must be at least 2″ in diameter.

The sample is mounted in the instrument together with a comparison surface (white chalk). The samples are placed in the instrument and light of different wavelength intervals is impinged on the surface. Reflected or transmitted light is then measured to obtain the property values. See Fig. 2−1.

Meaning of Data

This test is the primary method for obtaining colormetric data. The property determined that is of design interest is luminous transmittance.

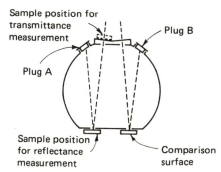

Fig. 2–1. Apparatus for measuring luminous transmittance.

B. MECHANICAL PROPERTIES

In the introductory remarks to the ASTM General Methods of Testing, it was implied that tests conducted on metals were, in a general way, adopted to plastics. This comment brings to mind the stress–strain curve of steel, which provides the essential characteristics of this metal. A glance at some of the plastic stress–strain curves shows a similarity to the curves for steel, but, when the plastics curve is superimposed on the steel curve, there is a pronounced difference between the two. The range of values for the plastic is so different from that of metals that it is hardly legible when shown on the same chart. (See Fig. 2–2.)

In addition to this, physical test results on plastics have only *short-term* significance. When a stress is applied over prolonged periods, the viscoelastic nature of the material will cause a permanent deformation coupled with a dimension change.

From these preliminary comments, it is evident that, in order to use the data sheet information, it is imperative to have a thorough understanding of how the data are evolved and what caution is to be exercised when applying the data to product design or other appropriate evaluations.

6. TENSILE PROPERTIES OF PLASTICS (ASTM-D638)

Specimen and Procedure

Specimens can be injection molded or machined from compression-molded plaques. They are given standard conditioning, and are typically ⅛″ thick, but can vary in this dimension. The shape is illustrated in Fig. 2–3.

Both ends of the specimen are firmly clamped in the jaws of an Instron testing machine. The jaws move apart at rates of 0.05, 0.2, or 20.0 inches per minute, pulling the specimen from both ends. The stress is automatically plotted against elongation (strain) on graph paper. When the speed of the

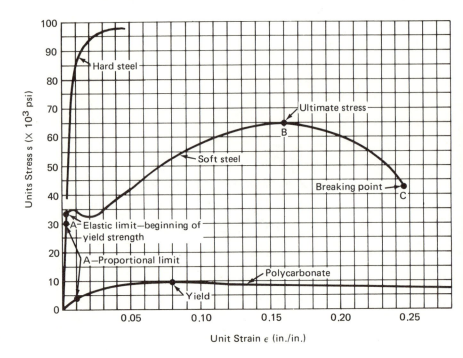

Unit Strain ε (in./in.)

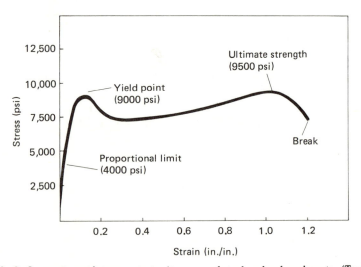

Strain (in./in.)

Fig. 2–2. Comparison of stress–strain diagrams of steel and polycarbonate. (Top) Polycarbonate stress–strain curve drawn to same scale as the steel curve. (Bottom) Polycarbonate curve drawn to a suitable scale.

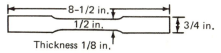

Fig. 2−3. Typical test specimen.

pulling jaws is not indicated on the data sheet, it is assumed that 0.2 inches per minute was used. The standard (data sheet) test is conducted at 73.4°C (23° C) and 50% relative humidity. A schematic outline of the test is shown in Fig. 2−4.

Meaning of Data

It should be recognized that tensile properties will most likely vary with a change of speed of the pulling jaws and with variation in the atmospheric conditions. Figure 2−5 shows the variation in a stress−strain curve when the speed of testing is altered; also shown are the effects of temperature changes

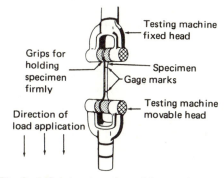

Fig. 2−4. Schematic outline of the tensile test.

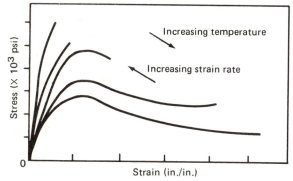

Fig. 2−5. Stress−strain curves showing the reaction of plastic to increasing temperatures and increasing rates of strain.

on the stress–strain curves. When the speed of pulling force is increased, the material reacts like brittle material; when the temperature is increased, the material reacts like ductile material.

The tensile data show the stress necessary to pull the specimen apart and the elongation prior to breaking. A moderate elongation (about 6%) of a test specimen implies that the material is capable of absorbing rapid impact and shock. The area under the stress–strain curve is indicative of overall toughness. A material of very high strength, high rigidity, and little elongation would tend to be brittle in service. For applications where almost rubbery elasticity is desirable, a high ultimate (over 100%) elongation may be an asset.

The tensile test and the calculated property data from it provide a most valuable source of information for the designer in determining product dimensions. The important consideration is that use conditions compare reasonably closely to the test conditions as far as speed of load, temperature, and moisture are concerned. Should use conditions differ appreciably, a test should be requested for data that are comparable to service requirements, which would thereby ensure that applicational needs will be based on more exacting data. The test is relatively inexpensive, and, where critical uses are encountered, it will eliminate interpolation and guessing. The tensile data are also useful for comparing various materials in this property.

It must be reemphasized that tensile data should only be applied to short-term stress conditions, such as operating a switch or shifting a clutch gear, etc.

The *yield point* is the first point on the stress–strain curve at which an increase in strain occurs without an increase in stress. The stress at which a material exhibits a special limiting deviation from the proportionality of stress to strain is the *yield strength*. A material whose stress–strain curve exhibits a point of zero slope may be considered to have a yield point. See Fig. 2–6.

Some materials exhibit a distinct discontinuity on the stress–strain curve within the elastic region. This is not a yield point by standard definition. Other materials do not exhibit a zero slope on the stress–strain curve, but display a gradual curvature in the yield region. In such cases the *offset yield strength* is used. It is obtained by establishing a point on the strain line where the strain exceeds, by a specified amount, the proportional limit. A line drawn through this newly established point parallel to the original proportional line when intersecting the curve will give the offset yield strength. See Fig. 2–7.

The data sheets usually omit the yield strength when there is no zero slope point on the stress–strain curve in the yield region.

In plastic materials, the values of the yield strength and the tensile strength are very close to each other.

The *tensile modulus (modulus of elasticity)* is another property derived from the stress–strain curve. Note that specimens used for this property test are separate from those used to determine tensile strength. The speed of testing— unless otherwise indicated—is 0.2 inches per minute, with the exception of

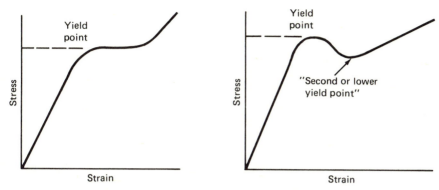

Fig. 2−6. Curves showing different yield points.

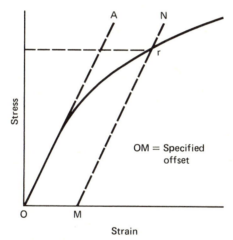

Fig. 2−7. Yield strength by offset method.

molded or laminated thermosetting materials in which the speed is 0.05 inches per minute. The tensile modulus is the ratio of stress to corresponding strain below the proportional limit of a material and is expressed in pounds per square inch.

The *proportional limit* is the greatest stress which a material is capable of sustaining without any deviation of the proportionality law. It is located on the stress−strain curve below the elastic limit.

The *elastic limit* is the greatest stress which a material is capable of sustaining without any permanent strain remaining upon complete release of the stress.

For materials that deviate from the proportionality law even well below the elastic limit, the slope of the tangent to the stress−strain curve at a low stress level is taken as the tensile modulus. When the stress−strain curve displays no proportionality at any stress level, the *secant modulus* is employed instead of the tensile modulus. The secant modulus is the ratio of stress to corresponding strain, usually at 1% strain.

The tensile modulus is an important property that provides the designer with information for a comparative evaluation of plastic material and also provides a basis for predicting the short-term behavior of a loaded part. Care must be used in applying the tensile modulus data to short-term loads to be sure that the conditions of the test are comparable to those in use. The longer-term modulus is treated under the ASTM-D674—Creep Test.

The tensile data can be applied to the design of short-term (3-hour duration) or intermittent loads in a product provided the use temperature, the humidity, and the speed of the load are within 10% of the test conditions outlined under the procedure. The intermittent specification merely indicates that there be sufficient time for strain recovery after the load has been removed.

The next step is to determine an allowable working stress. This is done by using a safety factor of $2\frac{1}{2}$ to 5 on the yield strength or tensile strength. If the type of stress is clearly defined, the $2\frac{1}{2}$ factor is adequate; otherwise, it should be higher. The final step is to calculate the elongation that the product would experience under the selected allowable working stress to see if such an elongation would permit the proper functioning of the product. The elongation could conceivably become the limiting component, and the working stress can be calculated from

$$\text{modulus} = \frac{\text{stress}}{\text{strain}} = E = \frac{S}{C}.$$

If product use conditions vary appreciably from those of the standard test, a stress–strain curve, derived using the procedure of anticipated requirement, should be requested and appropriate data developed.

7. FLEXURAL PROPERTIES OF PLASTICS (ASTM-D790)

Specimen and Procedure

Usually $\frac{1}{8}'' \times \frac{1}{2}'' \times 5''$ sheet or plaques as thin as $\frac{1}{16}''$ may be used. The span and width depend upon thickness. The specimen is placed on two supports spaced 4″ apart or 16−40 times the thickness, depending on the rigidity of specimen. A load is applied in the center at a specified rate, and the loading at failure (psi) is the flexural strength. The rate of strain is 0.1 in./in./min. For materials that do not break, the flexural property usually given is flexural stress at 5% strain.

Meaning of Data

These properties apply to parts subjected to bending, and, since many plastic products are involved in uses where bending stresses are generated, this test deserves close attention, especially in view of the viscoelastic nature of the materials.

It should be noted that test information will vary with *specimen thickness, temperature, atmospheric conditions, and a different speed of straining force.* This test is made at 73.4°F (23°C) and 50% relative humidity.

For brittle materials (those that will break below a 5% strain) the thickness, span, and width of the specimen and the speed of crosshead movement are varied to bring about a rate of strain of 0.01 in./in./min. The appropriate specimen size is shown in the ASTM table.

The *flexural strength* is the maximum stress that a material sustains at the moment of break. For materials that do not fail, the stress that corresponds to a strain of 5% is frequently reported as the flexural strength. See Fig. 2–8.

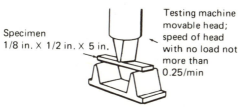

Fig. 2–8. Schematic outline of flexural strength test.

As a matter of interest it should be stated that in this test the force of bending and associated *deflection* are recorded. A formula gives the relationship between deflection and strain:

$$s = \frac{6dD}{L^2},$$

where

s = strain
d = depth of beam
D = deflection
L = span.

The *flexural yield strength* is determined from the calculated data of load-deflection curves that show a point where the load does not increase with an increase in deflection.

The *flexural modulus* is the ratio, within the elastic limit, of stress to corresponding strain. It is calculated by drawing a tangent to the steepest initial straight line portion of the load-deflection curve and using an appropriate formula.

In many plastic materials, as is the case with metals, when performing the flexural tests, increasing the speed of deflecting force makes the specimen appear more brittle and increasing the temperature makes it appear more ductile. This is the same relationship as in tensile testing. (See Fig. 2–5.)

When materials are evaluated against each other, the flexural data of those that break in the test can *not* be compared unless the conditions of the test and the specimen dimensions are identical. For those materials (most thermoplastics) whose flexural properties are calculated at 5% strain, the test conditions and the specimen are standardized and the data can be analyzed for relative preference. For design purposes, the flexural properties are used in the same way as the tensile properties. Thus, the allowable working stress, limits of elongation, etc., are treated in the same manner as are the tensile properties.

8. COMPRESSIVE PROPERTIES OF RIGID PLASTICS (ASTM-D695)

Specimen and Procedure

Prisms ½" × ½" × 1" or cylinders ½" diameter × 1" are used. The specimen is mounted in a compression tool between testing machine heads that move at a constant rate. The rate of movement is normally 0.05 inches per minute. An indicator registers loading. The test is made at 73.4°F (23°C) and 50% relative humidity.

The compressive strength of a material is calculated as the pounds per square inch required either to rupture the specimen or before the specimen deforms a given percentage of its height. It can be expressed as psi either at rupture or at a given percentage of deformation.

Meaning of Data

The compressive data are of limited design value. They can be used for comparative material evaluation and design purposes if the conditions of the test approximate those of application.

The data are of definite value for materials that fail in the compressive test by a shattering fracture. On the other hand, for those that do not fail in this manner, the compressive information is arbitrary and is determined by selecting a point of compressive deformation at which it is considered that a complete failure of the material has taken place. About 10% of deformation is viewed in most cases as maximum.

The test can provide compressive stress, compressive yield, and modulus. Many plastics do not show a true compressive modulus of elasticity. When loaded in compression, they display a deformation, but show almost no elastic portion on a stress–strain curve; those types of materials should be compressed with light loads. The data are derived in the same manner as in the tensile test.

9. SHEAR STRENGTH (ASTM-D732)

Specimen and Procedure

Sheets or molded discs from 0.0005" to 0.500" thick are used. The test is run at 73.4°F (23°C) and 50% relative humidity.

The specimen is mounted in a punch-type shear fixture, and the punch (1" diameter) is pushed down at a rate of 0.05 inches per minute until the moving portion of the sample clears the stationary portion.

Meaning of Data

Shear strength is calculated as the force per area sheared. Shear strength is particularly important in film and sheet products where failures from this type of load may often occur. This property can be used for comparison with other materials and for determination of the forces needed for punching openings (holes, etc.).

10. IZOD IMPACT (ASTM-D526)

Specimen and Procedure

The specimen is usually $\frac{1}{8}" \times \frac{1}{2}" \times 2"$. Specimens of other thicknesses can be used (up to $\frac{1}{2}"$), but $\frac{1}{8}"$ is frequently used for molding materials because it is representative of average part thickness.

A notch is cut on the narrow face of the specimen. (See Fig. 2−9.) The sample is clamped in the base of a pendulum testing machine (see Fig. 2−10) so that it is cantilevered upward with the notch facing the direction of impact. The pendulum is released, and the force expended in breaking the sample is calculated from the height the pendulum reaches on the follow-through. The speed of the pendulum at impact is 11 feet per second.

Meaning of Data

The Izod impact test indicates the energy required to break notched specimens under standard conditions. It is calculated as foot-pounds per inch of notch and is usually calculated on the basis of 1" specimen, although the specimen may be thinner in the lateral direction. (This is indicated in the sketch of the Izod specimen in Fig. 2−9.)

The Izod value is useful in comparing various types of grades of a plastic within the same material family. In comparing one plastic with another, however, the Izod impact test should not be considered a reliable indicator of overall toughness or impact strength. Some materials are notch-sensitive and develop greater concentrations of stress from the notching operation. It should be noted that the notch serves not only to concentrate the stress, but also to present plastic deformation during impact.

The Izod impact test may indicate the need to avoid inside sharp corners on parts made of such materials. For example, nylon and acetal-type plastics, which in molded parts are among the toughest materials, are notch-sensitive and register relatively low values on the notched Izod impact test.

See the *Tensile Impact* (ASTM-D1822) and the *Summary of "Impact Strength"* sections.

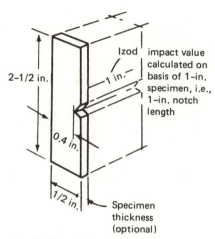

Fig. 2—9. Izod impact test specimen.

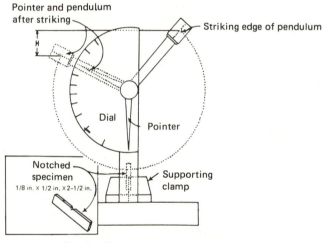

Fig. 2—10. Izod impact testing machine.

11. TENSILE IMPACT (ASTM-D1822)

Specimen and Procedure

Small and long specimens of "tensile bar" shape measuring 2.5″ and 6.35″ long, respectively, are the sizes of specimens, with the major change in dimensions in the necked-down section.

The specimen is mounted between a pendulum head and crosshead clamp on the pendulum of an impact tester. The pendulum is released and it swings past a fixed anvil which halts the crosshead clamp. The pendulum head continues forward, carrying the forward portion of the ruptured specimen.

The energy loss (tensile impact energy) is recorded, as well as whether the failure appeared to be of a brittle or ductile type. See Fig. 2−11.

Meaning of Data

This test is comparatively new. However, possible advantages over the notched Izod test are apparent: the notch sensitivity factor is eliminated, and energy is not used in pushing aside the broken portion of the specimen.

The test results are recorded in foot-pounds per square inch. This allows for minor variations in dimensions of the minimum in cross-section area.

Two specimens are used, S and L (short and long), so that the effect of elongation on the result can be observed. A ductile failure (best observed on the L specimen) results in a higher elongation and, consequently, in a higher total energy absorption than a brittle failure (best observed on the S specimen) in any specific material.

The energy for specimen fracture is a function of the force times the distance it travels. Thus two materials showing the same energy values in the tensile impact test (all elements of the test being the same) could consist of two different factors, such as a small force and a large elongation compared to a large force and a small elongation.

If one is to consider the application of these data to a design, the size of the force and its rate of application would have to be obtained and compared with the design requirement. The breakdown of energy into components of force and speed becomes possible by the addition of electronic instrumentation to the testing apparatus, thus enabling the supplier of the material to furnish additional information for material selection.

See the *Izod Impact* and *Summary of "Impact Strength"* sections.

Summary of "Impact Strength"

The data from the Izod and tensile impact tests can be comparatively evaluated especially when experience has been acquired with any one type of material. One should keep in mind the limitation mentioned in each *Meaning of Data* section.

Impact resistance is a significant characteristic of a material in many product designs. Associated with impact resistance is the term material toughness.

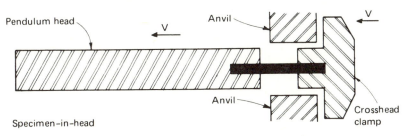

Fig. 2−11. Schematic outline of tensile impact machine.

Neither one can be measured in a way that is meaningful to the designer. The term "impact" implies a very high speed of the acting force, whereas "toughness" is not related to any specific speed. Since the two terms are used in conjunction with each other in describing resistance to impact, it appears desirable to correlate those readily obtainable properties that would reflect on speed of impact and toughness.

At a 73.4°F (23°C) test temperature and a speed of acting force listed in each test category, the following results prevail:

A high modulus and high Izod points to a very tough material

A high modulus and low Izod points to a brittle material

A low modulus and high Izod points to a flexible and ductile material

When use conditons differ from those applied to data sheet tests, the following comparative evaluation can be made:

Selecting an established high impact plastic-like polycarbonate as the standard, a tensile test would be made on this material at use speeds of striking force and end use environmental conditions. This provides a modulus and stress–strain curve. The same kind of test would be made on the materials being evaluated.

The area under a stress-strain curve is a measure of toughness. It thus becomes possible to compare the moduli and areas under the curve and thereby estimate impact strength as a percentage of the standard. The notch sensitivity factor is eliminated and a judgment element is introduced that can prove accurate if the information is diligently analyzed.

Where critical design areas involving safety to humans or protecting valuable devices are concerned, the simulation of end use with prototype parts (including extremes of conditions) is the most desirable way to test selected materials.

12. FATIGUE STRENGTH (ASTM-D671)

The fatigue strength in this specification is defined as that stress level at which the test specimen will sustain N cycles prior to failure. The data are generated on a machine which runs at 1800 cycles per minute. This test is of value to material manufacturers in determining consistency of their product.

For design purposes, these data could be useful only if loading, size and shape of part, frequency of stressing, and environmental conditions were exactly the same as service requirements. Such cases are next to impossible to find, therefore, the way to establish fatigue endurance is to test the part in question under use conditions.

13. ROCKWELL HARDNESS (ASTM-D785)
Specimen and Procedure

Sheets or plaques at least ¼″ thick are used. This thickness may be built up of thinner pieces, if necessary.

A steel ball under a minor load is applied to the surface of the specimen. This indents the specimen slightly and assures good contact. The gauge is then set at zero. The major load is applied for 15 seconds and removed, leaving the minor load still applied. The indentation remaining after 15 seconds is read directly off the dial.

The size of the balls used and loadings vary, and values obtained with one set cannot be correlated with values from another set. See Fig. 2−12.

Meaning of Data

Rockwell hardness can differentiate the relative hardness of different types of a given plastic; but, since elastic recovery is involved as well as hardness, it is not valid to compare the hardness of various types of plastic entirely on the basis of this test.

Hardness usually implies resistance to abrasion, wear, or indentation (penetration). In plastics it only means resistance to indentation. The scales range from

"R" with a major load of 60 kg; indenter of 0.5 in.

"L" with a major load of 60 kg; indenter of 0.25 in.

"M" with a major load of 100 kg; indenter of 0.25 in.

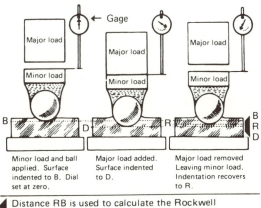

| Minor load and ball applied. Surface indented to B. Dial set at zero. | Major load added. Surface indented to D. | Major load removed Leaving minor load. Indentation recovers to R. |

Distance RB is used to calculate the Rockwell hardness figure, which can be read directly on the dial.

Fig. 2−12. Schematic outline of Rockwell hardness tester.

"E" with a major load of 100 kg; indenter of 0.125 in.

"K" with a major load of 150 kg; indenter of 0.125 in.

The hardness is of limited value to the designer, but can be of some value when comparing these data between materials.

14. DEFORMATION UNDER LOAD (ASTM-D621)

Specimen and Procedure

The specimen is ½" cube, either solid or composite. The specimen is placed between the anvils of the testing machine, and loaded at 1000, 2000, or 4000 psi. The gauge is read 10 seconds after loading, and again 24 hours later. The deflection is recorded in mils. The original height is calculated after the specimen is removed from the testing machine by adding the change in height to the height after testing. By dividing the change in height by the original height and multiplying by 100, the precentage deformation is calculated. This test may be run at 73.4, 122, or 158°F (23, 50, or 70°C). See Fig. 2–13.

Meaning of Data

This test on rigid plastics indicates their ability to withstand continuous short-term compression without yielding and loosening when fastened as in insulators or other assemblies by bolts, rivets, etc. It does not indicate the creep resistance of a particular plastic for long periods of time.

It is also a measure of rigidity at service temperatures and can be used as identification for procurement. Data should indicate stress level and the temperature of the test.

C. THERMAL PROPERTIES

15. DEFLECTION TEMPERATURE (ASTM-D648)

Specimen and Procedure

Specimens measure 5" × ½" × any thickness from ⅛" to ½". The specimen is placed on supports 4" apart and a load of 66 or 264 psi is placed on the center. The temperature in the chamber is raised at the rate of 2° ± 0.2° C per minute. The temperature at which the bar has deflected 0.010" is reported as "deflection temperature at 66 (or 264) psi fiber stress." See Fig. 2–14.

In this procedure, it is required to apply the load for 5 minutes and at the end of that period to set the measuring indicator to zero followed by initiation of the heating. The reason for the 5-minute delay is to compensate for the creep exhibited at room temperature by some plastics. At the 66-psi stress level the

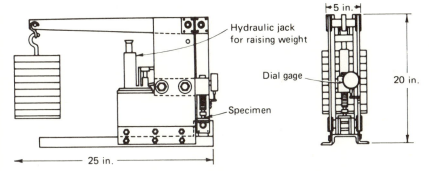

Fig. 2–13. Deformation testing machine.

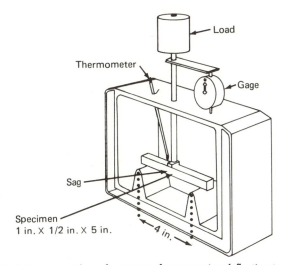

Fig. 2–14. Schematic outline of test setup for measuring deflection temperature.

amount of creep after 5 minutes (with most materials) will not constitute a significant percentage of the 0.01″ test deflection.

Meaning of Data

This test shows the temperature at which an arbitrary amount of deflection occurs under established loads. It is not intended to be a direct guide to high-temperature limits for specific applications, but may be useful in comparing the relative behavior of various materials in these test conditions. However, it is primarily useful for control and development purposes.

The deflection temperature is a valuable indicator of the increase in temperature resistance of filled and of reinforced materials. It is also used as a relative comparator of temperature characteristics between materials. The deflection temperature at 264 psi has been found for many materials to be the annealing temperature.

16. COEFFICIENT OF LINEAR THERMAL EXPANSION OF PLASTICS (ASTM-D696)

Specimen and Procedure

The specimen can be square or round to fit a ½″ I.D. dilatometer tube in a free sliding manner and is to be 2″–4″ long. The length is governed by the sensitivity of dial gauge, the expected expansion, and the accuracy desired.

The specimen is mounted in the dilatometer and placed in a bath of either −22°F (−30C°) or 187°F (+30°C) until the temperature of the bath is reached. When this takes place, the indicator dial is read showing the expansion or contraction of the specimen. These readings are compared with measurement of specimen length prior to placing it in the dilatometer. See Fig. 2–15.

Meaning of Data

With the increased application of plastics in combination with other materials, the coefficient of expansion plays an important role in making design allowances for expansions of various materials at different temperatures so that satisfactory functions of products are ensured.

The difference in thermal expansion between plastics and steel is very large. It is to be noted that some plastic material changes in length rather abruptly at some temperatures, beyond the limits of the test condition. In such cases, a special investigation should be instigated, and the coefficient of expansion established under temperatures of usage.

This test shows the reversible linear thermal expansion. The accuracy of these results may be affected by factors such as loss of plasticizer, solvent, relieving of stresses, etc., and, when a case demands most precise data, the factors mentioned should be considered for their possible influence on the information.

17. BRITTLENESS TEMPERATURE (ASTM-D746)

Specimen and Procedure

Pieces ½″ wide, 0.075″ thick, and 1¼″ long are the dimension of specimen.

The conditioned specimens are cantilevered from the sample holder in the test apparatus, which has been brought to a low temperature (that at which specimens would be expected to fail). When the specimens have been in the test medium for 3 minutes, a single impact is administered, and the samples are

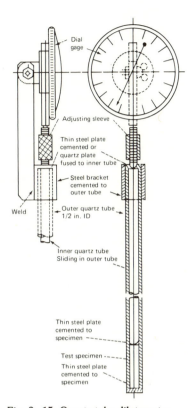

Fig. 2–15. Quartz tube dilatometer.

examined for failure. Failures are total breaks, partial breaks, or any visible cracks. The test is conducted at a range of temperatures producing varying percentages of breaks. From these data, the temperature at which 50% failure would occur is calculated or plotted and reported as the brittleness temperature of the material according to this test.

Meaning of Data

This test is of some use in judging the relative merits of various materials for low-temperature flexing or impact. However, it is specifically relevant only for materials and conditions specified in the test, and the values cannot be directly applied to other shapes and conditions.

The brittleness temperature does not put any lower limit on service temperature for end use products. The brittleness temperature is sometimes used in specifications.

18. FLAMMABILITY (FOR MATERIALS THICKER THAN 0.050") (ASTM-D635)

Specimen and Procedure

Specimens are $\frac{1}{2}''$ × 5" and "thickness normally supplied"—for thermoplastics this thickness is usually in the range of 0.05"–0.250". The specimens are tested in "as-received" condition, unless other conditioning is specified. Lines are scribed 1" from each end.

The specimen is clamped at one end on a ring stand so that the longitudinal axis is horizontal and the transverse axis is inclined 45° to the horizontal. A piece of 20-mesh Bunsen burner gauze is clamped horizontally $\frac{3}{8}''$ below the specimen. A Bunsen burner, placed so that the flame contacts the end of the specimen, is held 30 seconds and then removed. If the specimen does not ignite, the burner is returned for another 30-second attempt. The extent of burning is measured along the lower edge of the specimen.

Meaning of Data

If the specimen does not ignite, it is classed "nonburning by this test."

If the specimen continues to burn, it is timed until it stops or the 4-inch mark is reached. A specimen which burns to the 4-inch mark is classed as "burning by this test," and the rate is equal to (180/time) inches/minutes.

If the specimen does not continue burning to the 4-inch mark, it is classed as "self-extinguishing by this test," and the length of the burned portion is reported as the "extent of burning."

19. UNDERWRITERS LABORATORIES (UL) TEST 94

Specimen and Procedure

The specimen for the tests is 5" long × $\frac{1}{2}''$ wide × $\frac{1}{2}''$ maximum thickness. The preferred thickness is $\frac{1}{8}''$.

The placement of the specimen, the size of the flame, and its location with respect to the specimen are described in detail in the specifications.

For the horizontal burning test, the specimen is horizontally tilted at 45° and the lower edge is exposed to the prescribed location and size of flame. At the end of 30 seconds or prior to it, if the flame has reached 1", the burner, a Bunsen burner, is removed.

If the specimen continues to burn, the time to reach the 4-inch mark from the 1-inch mark on the specimen is recorded and the rate of burning is calculated.

For the vertical burning test, the specimen is clamped at one end and the opposite end is exposed to the prescribed location and size of flame for 10 seconds. The test flame is withdrawn and the duration of flaming is observed.

When specimen flaming ceases, the burner flame is immediately placed under the specimen.

After 10 seconds the flame is again withdrawn and flaming as well as glowing are noted.

Meaning of Data

The purpose of flammability tests is to measure and describe the flammability properties of materials used in devices and appliances in response to heat and flame under controlled laboratory conditions.

Thus the materials tested under the horizontal burning test 94HB shall conform to the following:

A. Not have a burning rate exceeding 1.5 inches per minute over 3.0-inch span for specimens tested in a thickness of 0.120−0.500 inch; or

B. Not have a burning rate exceeding 3.0 inches per minute over 3.0-inch span for specimens having a thickness less than 0.120 inch;

C. Cease to burn before the flame reaches the 4.0-inch reference mark.

The materials tested under the vertical burning test 94V shall conform to the following:

Materials Classed 94V-0

Materials classed 94V-0 shall:

A. Not have any specimens which burn with flaming combustion for more than 10 seconds after either application of the test flame.

B. Not have a total flaming combustion time exceeding 50 seconds for the 10 flame applications for each set of five specimens.

C. Not have any specimens which burn with flaming or glowing combustion up to the holding clamp.

D. Not have any specimens which drip flaming particles that ignite the dry absorbent surgical cotton located 12 inches below the test specimen.

E. Not have any specimens with glowing combustion which persists for more than 30 seconds after the second removal of the test flame.

Materials Classed 94V-1

Materials classed 94V-1 shall:

A. Not have any specimens which burn with flaming combustion for more than 30 seconds after either application of the test flame.

B. Not have a total flaming combustion time exceeding 250 seconds for the 10 flame applications for each set of five specimens.

C. Not have any specimens which burn with flaming or glowing combustion up to the holding clamp.

D. Not have any specimens which drip flaming particles that ignite the dry absorbent surgical cotton located 12 inches below the test specimen.

E. Not have any specimens with glowing combustion which persists for more than 60 seconds after the second removal of the test flame.

Materials Classes 94V-2

Materials classed 94V-2 shall:

A. Not have any specimens which burn with flaming combustion for more than 30 seconds after either application of the test flame.

B. Not have a total flaming combustion time exceeding 250 seconds for the 10 flame applications for each set of five specimens.

C. Not have any specimens which burn with flaming or glowing combustion up to the holding clamp.

D. Be permitted to have specimens that drip flaming particles which burn only briefly, some of which ignite the dry absorbent surgical cotton placed 12 inches below the test specimen.

E. Not have any specimens with glowing combustion which persists for more than 60 seconds after the second removal of the test.

Vertical Burning Test for Classifying Materials 94-5V

Materials classed 94-5V shall:

A. Not have any specimens which burn with flaming and/or glowing combustion for more than 60 seconds after the fifth flame.

B. Not have any specimens which drip any particles.

20. OXYGEN INDEX FOR PLASTICS (ASTM-D2863)

The test method describes a procedure for measuring the minimum concentration of oxygen in a flowing mixture of oxygen and nitrogen that will support glowing combustion of plastics.

The oxygen index is the minimum concentration of oxygen expressed as a volume percent in a mixture of oxygen and nitrogen that will just support glowing combustion of a material initially at room temperature under the conditions of this method.

From this description, it is apparent that the lower the oxygen index the more the plastic contributes to the support of combustion.

Note: The three preceding property data relate to flammability of plastics. Many of the basic polymers require additives that will improve their resistance to supporting combustion.

These improvements vary in degree, and the designer must be cautioned not to overspecify the requirement for flammability. It should be recognized that the highest protection against burning can be costly, can contribute to higher specific gravity, and can adversely affect mechanical properties. The flammability resistance should be just high enough to meet product requirements.

21. DIELECTRIC STRENGTH (ASTM-D149)

Specimen and Procedure

Specimens are thin sheets or plates having parallel plane surfaces and are of a size sufficient to prevent flashing over. Dielectric strength varies with thickness and, therefore, specimen thickness must be reported.

Since temperature and humidity affect results, it is necessary to condition each type of material as directed in the specification for that material. The test for dielectric strength must be run in the conditioning chamber or immediately after removing the specimen from the chamber.

The specimen is placed between heavy cylindrical brass electrodes, which carry electrical current during the test. There are two ways of running this test for dielectric strength:

1. *Short-Time Test:*

   The voltage is increased from zero to breakdown at a uniform rate—0.5 – 1.0 kV/sec. The precise rate of voltage rise is specified in the governing material specifications.

2. *Step-By-Step Test:*

   The initial voltage applied is 50% of breakdown voltage shown by the short-time test. It is increased at rates specified for each type of material, and the breakdown level is noted.

Breakdown by these tests means passage of sudden excessive current through the specimen and can be verified by instruments and by visible damage to the specimen.

Meaning of Data

This test is an indication of the electrical strength of a material as an insulator. The dielectric strength of an insulating material is the voltage gradient at which electric failure or breakdown occurs as a continuous arc (the electrical property analogous to tensile strength in mechanical properties). The dielectric strength of materials varies greatly with several conditions, such as humidity and geometry, and it is not possible to directly apply the standard test values to field use unless all conditions, including specimen dimensions, are the same. Because of this, the dielectric strength test results are of relative rather than absolute value as a specification guide.

The dielectric strength of polyethylenes is usually around 500 V/mil. The value will drop sharply if holes, bubbles, or contaminants are present in the specimen being tested.

The dielectric strength varies inversely with the thickness of the specimen.

22. DIELECTRIC CONSTANT AND DISSIPATION FACTOR (ASTM-D150)

Specimen and Procedure

The specimen may be a sheet of any size convenient to test, but should have uniform thickness. The test may be run at standard room temperature and humidity, or in special sets of conditions as desired. In any case, the specimens should be preconditioned to the set of conditions used.

Electrodes are applied to opposite faces of the test specimen. The capacitance and dielectric loss are then measured by comparison or substitution methods in an electric bridge circuit. From these measurements and the dimensions of the specimen, dielectric constant and loss factor are computed.

Meaning of Data

The dissipation factor is a ratio of the real power (in-phase power) to the reactive power (power 90° out of phase). It is also defined as:

The dissipation factor is the ratio of conductance of a capacitor in which the material is the dielectric to its susceptance.

The dissipation factor is the ratio of its parallel reactance to its parallel resistance. It is the tangent of the loss angle and the cotangent of the phase angle.

The dissipation factor is a measure of the conversion of the reactive power to real power, showing as heat.

The *dielectric constant* is the ratio of the capacity of a condenser made with a particular dielectric to the capacity of the same condenser with air as the dielectric. For a material used to support and insulate components of an electrical network from each other and ground, it is generally desirable to have a low level of dielectric constant. For a material to function as the dielectric of a capacitor, on the other hand, it is desirable to have a high value of dielectric constant, so that the capacitor may be physically as small as possible.

The *loss factor* is the product of the dielectric constant and the power factor, and is a measure of total losses in the dielectric material.

23. TESTS FOR ELECTRICAL RESISTANCE (ASTM-D257)

Specimen and Procedure

Specimens for these tests may be any practical form, such as flat plates, sheets, or tubes.

These tests describe methods for determining the several properties defined below. Two electrodes are placed on or embedded in the surface of a test specimen. The following properties are calculated.

Property: Insulation Resistance

Definition:

Ratio of direct voltage applied to the electrodes to the total current between them; dependent upon both volume and surface resistance of the specimen.

Meaning of Data:

In materials used to insulate and support components of an electrical network, it is generally desirable to have insulation resistance as high as possible.

Property: Volume Resistivity

Definition:

Ratio of the potential gradient parallel to the current density.

Property: Surface Resistivity

Definition:

Ratio of the potential gradient parallel to the current along its surface to the current per unit width of the surface.

Meaning of Data:

Knowing the volume and surface resistivity of an insulating material makes it possible to design an insulator for a specific application.

Property: Volume Resistance

Definition:

Ratio of direct voltage applied to the electrodes to that portion of current between them that is distributed through the volume of the specimen.

Property: Surface Resistance

Definition:

Ratio of the direct voltage applied to the electrodes to that portion of the current between them that is in a thin layer of moisture or other semiconducting material which may be deposited on the surface.

Meaning of Data:

High volume and surface resistance are desirable in order to limit the current leakage of the conductor that is being insulated.

24. ARC RESISTANCE (ASTM-D495)

Meaning of Data

This test shows the ability of a material to resist the action of an arc of high voltage and low current close to the surface of the insulation in tending to form a conducting path therein.

The arc resistance data are of relative value only for distinguishing materials of nearly identical composition, such as for quality control, development, or identification.

3. Description and Derivation of Long-Term Properties

In Chapter 2 the data sheet short-term tests were described and explained. In this chapter we shall concern ourselves with the long-term behavior of plastics when exposed continuously to stresses, environment, excessive heat, abrasion, and continuous contact with liquids.

Tests outlined by ASTM and described in detail are intended to produce consistency in observations and records by various manufacturers, so that they can be correlated to provide meaningful information to product designers.

1. LONG TIME CREEP AND STRESS— RELAXATION OF PLASTICS (ASTM-D2990)

Specimen and Procedure

The procedure under this heading is intended as a recommendation for uniformity of making setup conditions for the test, as well as recording the resulting data. The reason for this move is the time-consuming nature of the test (many years duration), which does not lend itself to routine testing.

The test specimen can be round, square, or rectangular and manufactured in any suitable manner. For round specimens, an 0.505-inch, 0.357-inch, or 0.252-inch diameter with a gauge length of 2 inches is recommended. The test is conducted under controlled temperature and atmospheric conditions.

The specimens for flexural creep can be the same as called for in tests for "Flexure."

The requirements for consistent results are outlined in detail as far as accuracy of time interval, of readings, etc., in the ASTM procedure. Each report of test results should indicate the exact grade of material and its supplier, the specimen's method of manufacture, its original dimensions, type of test (tension, compression, or flexure), temperature of test, stress level, and interval of readings.

Meaning of Data

Creep at room temperature is called *cold flow*. When a load is initially applied to a specimen, there is an instantaneous strain or elongation. Subsequent to this, there is the time-dependent part of the strain called *creep*, which results from the continuation of the constant stress at a constant temperature. In terms of design, creep means changing dimensions and deterioration of product strength when the product is subjected to a steady load over a prolonged period of time.

All the mechanical properties described in tests for the data sheet properties represented values of short-term application of forces, and, in most cases, the data obtained from such tests were used for comparative evaluation or as controlling specifications for quality determination of materials along with short-duration and intermittent-use design requirements.

A stress–strain diagram is a significant source of data for a material. In metals, for example, most of the needed data for mechanical property considerations are obtained from a stress–strain diagram, e.g., Fig. 2–2. In plastic, however, the viscoelasticity causes an initial deformation at a specific load and temperature and is followed by a continuous increase in strain under identical test conditions until the product is either dimensionally out of tolerance or fails in rupture as a result of excessive deformation. This type of an occurrence can be explained with the aid of a Maxwell model, shown in Fig. 3–1.

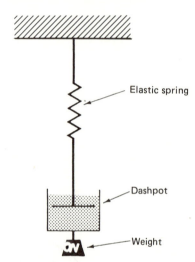

Fig. 3–1. Maxwell model to explain viscoelastic behavior.

With a load applied to the system, shown diagrammatically, the spring will deform to a certain degree. The dashpot will at first remain in a stationary position under the applied load, and, if the same load continues to be applied, the viscous fluid will slowly leak past the piston which will cause the dashpot to move. The dashpot movement corresponds to the strain or deformation of the plastic material.

When the stress is removed, the dashpot does not return to its original position, as does the spring. Thus we can visualize a viscoelastic material as having dual actions: one of an elastic material like the spring and the other like the viscous liquid in the dashpot. The properties of the elastic phase are independent of time; however, the properties of the viscous phase are very much a function of time, temperature, and stress. The phenomenon is further explained by looking at the dashpot again where we can visualize that a thinner fluid resulting from increased temperature under a higher pressure (stress) will have a higher rate of leakage around the piston during the time that the above conditions prevail. Translated into plastic creep, it means that at higher use temperature and higher stress levels the strain will be higher, therefore resulting in greater creep.

The visualization of the reaction to a load by the dual component interpretation of a material is valuable to the understanding of the creep process, but meaningless for design purposes. For this reason, the designer is interested in actual deformation or part failure over a specific time span. This means making observations of the amount of strain at certain time intervals which will make it possible to construct curves that could be extrapolated to longer time periods. The initial readings are 1, 2, 3, 5, 7, 10, and 20 hours, followed by readings every 24 hours up to 500 hours and then readings every 48 hours up to 1000 hours.

The time segment of the creep test is common to all materials, i.e., strains are recorded until the specimen ruptures or the specimen is no longer useful because of yielding. In either case, a point of failure of the test specimen has been reached.

The stress levels and the temperature of the test for a material is determined by the manufacturer. The guiding determinants are the continuous allowable working stress at room temperature and the continuous allowable working stress at temperatures of potential applications.

The strain readings of a creep test can be more convenient to a designer if they are presented as a creep modulus. In a viscoelastic material, strain continues to increase with time while the stress level remains constant. Since the modulus equals stress divided by strain, we have the appearance of a changing modulus.

The *creep modulus*, also known as *apparent modulus* or *viscous modulus* when graphed on log-log paper, is normally a straight line and lends itself to extrapolation for longer periods of time. The apparent modulus should be differentiated from the modulus given in the data sheets, because the latter is an instantaneous value derived from the testing machine.

The method of obtaining creep data and their presentation have been described; however, their application is limited to the exact same material, temperature use, stress level, atmospheric conditions, and type of test (tensile, compression, flexure) with a tolerance of ±10%. Only rarely do product requirement conditions coincide with those of the test or, for that matter, are creep data available for all grades of material that may be selected by a designer. In those cases a creep test of relatively short duration—1000 hours—can be instigated, and the information can be extrapolated to the long-term needs.

It should be noted that reinforced thermoplastics and thermosets display much higher resistance to creep.

There have been numerous attempts to develop formulas that could be used to predict creep information under varying usage conditions. In practically all cases, suggestions are made to verify the calculated data with actual test performance. Furthermore, the introduction of numerous factors that have to be developed for specific materials and usage conditions make it cumbersome to apply such means for reliable predictions of product behavior.

The method of applying creep data will be discussed further in Chapter 7. The data in Fig. 3−2 have been plotted from available and published information of material manufacturers. The first point is the 100-hour time interval. Up to that point, the data for the shorter intervals do not, as a rule, fit the straight-line configuration that exists on log-log charts for the long-term duration beyond the first 100-hour test period. The circled points are the 100-, 300-, and 1000-hour test periods, and a straight line is fitted either through the circles or tangent to them to give the line a slope for long-term evaluation.

From this line, we can estimate at what time the strain will be such that it will cause tolerance problems in the product performance. Or if we use the elongation at yield as the point at which the material has attained the limit of its useful life, we can estimate the time at which this is reached.

The formula modulus (apparent) = stress/strain enables us to locate the modulus that corresponds to the test stress and strain, obtained by limiting the dimensional change or the elongation limit, where it intersects the straight line leading to an appropriate time value. For example, the polycarbonate creep line shows that a limit of 0.010 in elongation is reached at the end 10^5 hours and the elongation of 0.06 (yield) is arrived at after 10^{17} hours, or indefinitely if the 0.010 limitation does not exist.

Creep information is not as readily available as short-term property data sheets are. From a designer's viewpoint, it is important to have creep data available for products subjected to a constant load for prolonged periods of time. Even if standard test creep data are available, the chances are that the conditions of the test will not reasonably correspond to those of the product use that is contemplated, such as stress level, temperature, or environmental surroundings. In the interest of sound designing procedure, the necessary creep information should be procured on the prospective material and under conditions of product usage. In addition to the creep data, a stress−strain diagram, also at conditions of product usage, should be obtained. The combined information will provide the basis for calculations and the predictability of material performance.

The needed data can be requested from the material supplier or, if that is not feasible within the time frame of the designer's needs, an independent laboratory could be asked to perform the test, and, finally, as an outside possibility, the test could be carried out by personnel of the designing group.

What we are talking about in terms of time is a reading after initiation of the test at 4 days, 8 days later, and finally at the end of 42 days.

The cost of making the test in comparison with other expenditures related to product design would be insignificant when considering the element of safety and confidence it would provide. Furthermore, the proving of product performance could be carried out with a higher degree of favorable expectations as far as plastic material is concerned.

Progressive material manufacturers can be expected to supply the needed creep and stress–strain data under specified use conditions when requested by the designer; but, if that is not the case, other means should be utilized to obtain the required information.

In conclusion, it can be stated that creep data and a stress–strain diagram will indicate whether plain resin properties can lead to practical product dimensions, or whether a reinforced resin has to be substituted to keep the design within the desired proportions.

For long-term product use under continuous load, plastic materials have to be considered with much greater care than would be the case with metals.

Stress Relaxation

When a part is deliberately deformed to a predetermined dimension and shape and is constrained in that position so that further deformation will neither increase nor decrease, we have a phenomenon known as *stress relaxation*. In such a case the initial resistance to deformation gradually decreases and can reach zero if the part is left in the restrained position long enough. At this time, upon removal of the deforming means (force) the part would retain the shape into which it was deformed.

Creep and stress relaxation are occurrences of the same type, i.e., in both cases permanent deformation takes place.

Stress relaxation curves can be constructed from observations of the remaining stresses in a specimen that is constrained under constant temperature as well as environment, and stressed in a manner described in the introductory comment.

2. PERMANENT EFFECT OF HEAT (ASTM-D794)

Specimen and Procedure

The specimen may be any piece of plastic or molded plastic part. It is placed in an air-circulating oven at a temperature (a multiple of 25°C) that is thought or known to be near the temperature limit of the material. If, after 4 hours, no change is observed, the temperature is increased in increments of 25°C at 4-hour intervals until a change does occur.

The change might be any property or properties of special interest— mechanical, visual, dimensional, color, etc. The test is written so that many effects of heat can be studied and specification requirements can be individually agreed upon by the parties concerned.

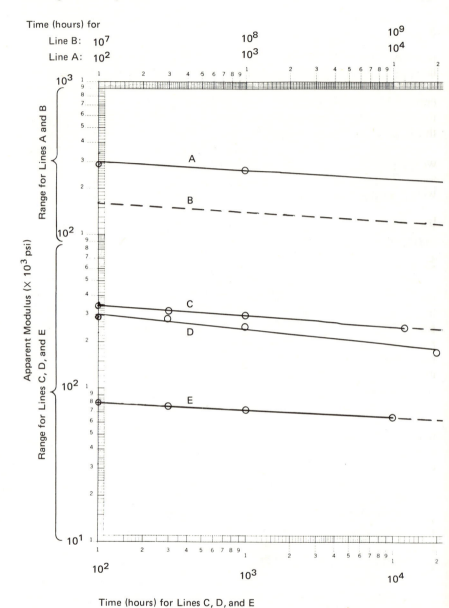

Fig. 3–2. Apparent modulus versus time. (A) Merlon, polycarbonate, 2000 psi at 73°F (23°C). (B) Extrapolation of (A) beyond 10⁷ hours. (C) Noryl 731, modified PPO, 2000 psi at 73°F (23°C). (D) Delrin 500, acetal, 1000 psi at 73°F (23°C). (E) Zytel 109, nylon, 50% relative

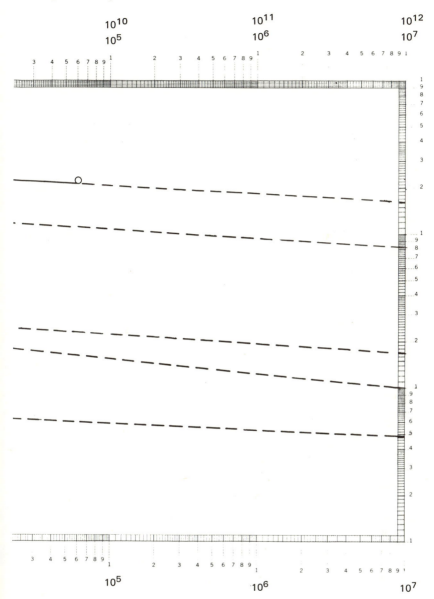

Fig. 3-2 (continued). humidity and 1000 psi at 73°F (23°C). The broken lines represent extrapolated values. The circles are actual test reading points.

Note: The log-log graph sheets are 9 in. × 15 in. and contain 3 × 5 cycles. The end of the first time cycle represents 1000 hr (10^3 hr). The creep graphs in Chapter 5 are reduced in scale. The lines in each cycle correspond to the numbered lines within a cycle on this sheet.

Meaning of Data

This test is of particular value in connection with established or potential applications that involve service at elevated temperatures. It permits comparison of various plastics and grades of one plastic in the form of test specimens, as well as molded parts in finished form.

3. ACCELERATED WEATHERING (ASTM-E42)

Specimen and Procedure

The specimen may be any shape, with size up to 5″ × 7″ × 2″. Artificial weathering has been defined by ASTM as "The exposure of plastics to cyclic laboratory conditions involving changes in temperature, relative humidity, and ultraviolet (UV) radiant energy, with or without direct water spray, in an attempt to produce changes in the material similar to those observed after long-term continuous outdoor exposure."

Three types of light sources for artificial weathering are in common use:

Source	UV Energy Output, Approx. (× Sunlight)
Enclosed UV carbon arc	7.5
Open-flame sunshine carbon arc	3
Water-cooled xenon arc	1

Selection of the light source involves many conditions and circumstances, such as the material is being tested, the proposed end-use, previous testing experience, or the type of information desired.

Meaning of Data

Since weather varies from day to day, year to year, and place to place, no precise correlation exists between artificial laboratory weathering and natural outdoor weathering. However, standard laboratory test conditions produce results with acceptable reproducibility and in general agreement with data obtained from outdoor exposures. Fairly rapid indications of weatherability are therefore obtainable on samples of known materials that through testing experience over a period of time have general correlations established. There is no artificial substitute for precisely predicting outdoor weatherability on materials with no previous weathering history. Weatherometers produce conditions to accelerate effects that would be observed in specimens exposed outdoors.

Accelerated Exposure to Sunlight Using the Atlas Type FDA−R

The Atlas Type FDA−R Fadeometer[1] is used primarily to check and compare color stability. Besides determining the stability of various pigments

[1] There is no standard ASTM test for the use of this equipment.

needed to provide both standard and custom colors, the Fadeometer is helpful in preliminary studies of various stabilizers, dyes, and pigments compounded in plastics to prolong their useful life. It is primarily for testing materials to be used in articles subject to indoor exposure and to sunlight.

The Fadeometer was used extensively in the development of UV absorbing acetate film for store windows to protect merchandise displayed in direct sunlight.

Exposure in the Fadeometer cannot be related directly to exposure in direct sunlight, partially because other weather factors are always present outdoors.

4. OUTDOOR WEATHERING (ASTM-D1435)

Specimen and Procedure

The specimen has no specified size. Specimens for this test may consist of any standard molded test specimen or cut pieces of sheet or machined samples.

Specimens are mounted outdoors on racks slanted at 45° and facing south. It is recommended that concurrent exposure be carried out in many varied climates to obtain the broadest, most representative total body of data. Sample specimens are kept indoors as controls and for comparison.

Reports of weathering describe all changes noted, areas of exposure, and period of time.

Meaning of Data

Outdoor testing is the most accurate method of obtaining a true picture of weather resistance. The only drawback of this test is the time required for several years exposure.

A large number of specimens are usually required to allow periodic removal and to run representative laboratory tests after exposure.

5. WEIGHT LOSS ON HEATING (IN ASTM-D706-63—SPECIFICATIONS FOR CELLULOSE ACETATE MOLDING COMPOUNDS)

Specimen and Procedure

Specimens measure $3'' \times 1'' \times \frac{1}{8}''$ conditioned 48 hours over anhydrous calcium chloride at 73.4°F (23°C).

Conditioned specimens are weighed and placed in an oven for 72 hours at 180°F (82°C). When removed from the oven, they are cooled in a desiccator over anhydrous calcium chloride to 73.4°F (23°C). The specimens are then weighed, and the percentage of loss is calculated.

Meaning of Data

This test for cellulosics indicates general aging stability, especially when service at elevated temperatures is anticipated. It quickly shows, in weight loss, dimensional change, or both, whether a particular formulation appears sufficiently stable for a given application.

Plasticizer volatility is generally the cause of weight loss on heating, and this test is commonly used to help select plasticizers that give a desired balance of properties and stability.

6. ENVIRONMENTAL STRESS CRACKING (ASTM-D1693-60T)

Specimen and Procedure

This test is limited to type I (low-density) polyethylenes. Specimens measure $\frac{1}{8}'' \times \frac{1}{2}'' \times 1''$. These are annealed in water or steam at 212°F (100°C) for 1 hour and then equilibrated at room temperature for 5–24 hours. After conditioning the specimens are nicked according to directions given.

The specimens are bent into a U shape in a brass channel and inserted into a test tube that is then filled with fresh reagent (Igepal). The tube is stoppered with an aluminum-covered cork and placed in a constant temperature bath at 122°F (50°C). These are inspected periodically and any visible crack is considered a failure. The duration of the test is reported along with the percentage of failures.

Meaning of Data

The cracking obtained in this test is indicative of what may be expected from a wide variety of stress-cracking agents. The information cannot be translated directly into end-use service prediction, but serves to rank various types and grades of polyethylene categories of resistance to environmental stress cracking.

Though restricted to type I polyethylene, this test can be used on high- and medium-density materials as well, in which case it would be considered a "modified" test.

7. ABRASION AND MAR RESISTANCE (ASTM-D1044, D1242, AND D673)

Procedure

In the D1044 test, the loss of optical effects is measured when a specimen is exposed to the action of a special abrading wheel.

In the D1242 test, the amount of material lost by a specimen is determined when the specimen is exposed to falling abrasive particles or to the action of an abrasive belt.

In the D673 test, the loss of gloss due to the dropping of loose abrasive on the specimen is measured.

Meaning of Data

The results produced by the three different tests may be of value for research and development work when it is desired to improve a material with respect to one of the test methods.

The variables that enter into tests of this type are so numerous that it is questionable how the information from such tests could be usable. Some of the factors are type of abrasive, shape of abrading particle, nature of plastic material, speed of action on the plastic, shape of the part and its temperature, the manner in which the abrasive is attached to the backing, and the bonding agent.

Currently, these tests are of no practical value to the designer and the only approach to the problem of scratch, mar, and abrasion resistance is to simulate actual performance needs.

For optical purposes, a cast sheet in the allyl family of plastics known as CR39 manufactured by PPG Co. has been used as a standard of comparison in evaluating scratch and mar resistance of a material. The CR39 is used for eye lenses and other optical products where the advantages of plastics are a consideration.

In recent years, coatings have been developed for polycarbonate and acrylics that dramatically improve the scratch and mar resistance of these materials.

8. WATER VAPOR PERMEABILITY (ASTM-E96)

Procedure

The material to be tested is fastened over the mouth of a dish that contains either water or a desiccant. This assembly is placed in an environment of constant humidity and temperature. The gain or loss in weight of the assembly is used to calculate the rate of water vapor movement through the specimen under prescribed conditions of humidity inside and outside of the dish. The results are reported in grams per 100 square inches during 24 hours, or equivalent metric units.

Meaning of Data

It should be recognized that all plastic materials allow a certain amount of water vapor, organic gas, or liquid to permeate the thickness of the material—it is only a matter of degree of permeation between various materials used as barriers against vapors and gases.

It has been found that the *permeability coefficient* is a function of the *solubility coefficient* and *diffusion coefficient*. The process of permeation is explained as the solution of the vapor into the incoming surface of the barrier, followed by diffusion through the barrier thickness, and evaporation on the exit side.

The problem of permeability exists whenever a plastic material is exposed to vapor, moisture, or liquids. Typical cases are electrical batteries, instrumentation, components installed underground, encapsulated electrical components, food packaging, and various fluid-material containers. In these cases, a plastic material is called upon to form a barrier either to minimize loss of vapor or fluid or to prevent the entrance of vapor or fluid into a part.

The factors influencing permeation are

1. The composition of the barrier, including additives, fillers, colorants, plasticizers, etc. Even when data on permeability are available for the basic polymer, they cannot be used for evaluation because commercial grades contain some ingredients that can change the values.

2. Crystalline polymers are better vapor barriers than amorphous polymers. Also, thermosetting resins have better barrier properties, especially when the fillers are nonmoisture absorbing.

3. An increase in temperature brings about an increase in permeability. Additionally, an increase in vapor pressure of the permeating agent also causes acceleration of transmission.

4. Part thickness is inversely proportional to permeation, i.e., with double the thickness, there is one-half of the evaporation.

5. Coatings such as epoxy-based finishes will improve resistance to permeation.

6. In the case of organic vapors, the permeation will depend not only on the composition of the barrier, but also on the molecular configuration of both the barrier material and the permeating agent.

Summary

From the designers' viewpoint, the tolerable amount of permeation established by test under conditions of usage with a part of correct shape and material is the only direct answer.

9. CONDITIONING PROCEDURES (ASTM-D618)

Procedure

Procedure A for conditioning test specimens calls for the following periods in standard laboratory atmosphere [50 ± 2% relative humidity, 73.4 ± 1.8°F (23 ± 1°C)]:

Specimen Thickness (inch)	Time (hour)
0.25 or under	40
Over 0.25	88

Adequate air circulation around all specimens must be provided.

Reason For Test

The temperature and moisture content of plastics affects the physical and electrical properties. In order to get comparable test results at different times and in different laboratories a standard has been established.

In addition to Procedure A described above, there are other conditions set forth to provide for testing at higher or lower levels of temperature and humidity.

All specimens are normally conditioned to ASTM-D618.

4. Polymer Formation and Variation

INTRODUCTION

Polymer formation is a complex subject, whose proper treatment would require hundreds of pages of theory and the application of polymer chemistry. This chapter is intended to serve only as a refresher on how polymers are formed. The nature of this volume limits the description to a condensed outline that will provide a general understanding of how plastics are created.

In the process of selecting a plastic material for a product, chemical terms are frequently used that may have some bearing on material properties or their processing, but may not be understood fully by those who are responsible for the performance of a product. It will be the aim in the following descriptions, not only to use such terms, but also to explain them for designers.

POLYMER FORMATION

Briefly stated, the creation of polymers can be explained as follows: Atoms are joined to form molecules. The molecules are caused to combine with each other in a long-chain form. These chains fold, intertwine with each other, and hold together with interchain forces. This molecular chain mixture results in a plastic raw material. We will now elaborate on these steps and see how we can visualize the details of creating a plastic raw material.

The starting point of a chemical compound is the atom, which is the reacting unit that takes part in a chemical change. To see how atoms react and join to initiate chemical compounds, we will examine the structure of the atom.

The hydrogen atom is the smallest and simplest atom. (See Fig. 4−1.) It is visualized as consisting of a single positively charged particle—a *proton*—which is the nucleus, and a single negatively charged particle—an *electron*—outside the nucleus. The electron is very much smaller (1 to 2000 in mass) than the proton and is visualized as orbiting the proton in the same way as a planet revolves around the sun. The positively charged proton and negatively charged electron produce an electrically neutral hydrogen atom.

Each atom has its own structure in terms of protons and electrons, as, for example, C and Cl (carbon and chlorine). (See Fig. 4−2.) The number of protons indicated in the nucleus is called the atomic number. In addition to the protons, the nucleus consists of *neutrons*, which are neutral building units with a mass almost equal to that of the protons, but slightly heavier to balance the positive charge in the nucleus. The number of neutrons does not necessarily correspond to the number of protons.

The conceptual arrangement of electrons for chlorine shows an inner orbit of two electrons, which act to stabilize the nucleus. This is common in all elements. The next orbit contains eight electrons, and, finally, the outer shell

Hydrogen atom

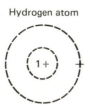

Fig. 4–1. Schematic of the hydrogen atom. The orbit of the electron is represented by the dotted line, the nucleus is represented by the heavy line, and the charged particles are represented by + and − signs.

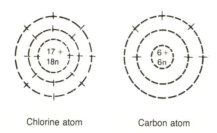

Chlorine atom   Carbon atom

Fig. 4–2. Schematic of the chlorine and carbon atoms. *Chlorine* (Cl) has seven electrons in the outer shell and would gain in symmetry if it picked up one electron giving it a valence of −1. *Carbon* (C) has four electrons in the outer shell and could either lose four electrons or pick up four electrons giving it a valence of plus or minus 4.

contains seven electrons, which would gain in symmetry if it picked up one additional electron. According to the octet rule, eight electrons provide a balanced condition. This rule applies to atoms below the third period of the periodic table. The *periodic table* is a placement of elements in a certain order in which some characteristics of elements tend to periodically repeat themselves.

The next step in our discussion is to learn how atoms combine. One of the ways in which some atoms are joined or bonded together can be explained by the experiment shown in Fig. 4–3. A beaker is filled with a solution of hydrochloric acid (HCl) and two electrodes, placed in the solution, are connected to a storage battery. After closing the switch in the circuit, we observe, with the ammeter, a current flowing, and, after a short period, we note that gas bubbles are gathering on each electrode. If the gas bubbles were to be examined, we would find hydrogen collected on the negative electrode and chlorine on the positive. What is happening is the splitting of the HCl molecule into charged atoms known as *ions*, and *not* into electrically neutral atoms. In order for the atoms to be charged, an internal change has to take place. The hydrogen atom, having the outer electron (negative) revolving in its orbit, will lose it to the chloride atom, which will then be negatively charged and will be attracted by the positive electrode. This leaves the nucleus of the hydrogen atom with a positive charge (proton) to be attracted by the negative electrode. On contact between the hydrogen proton and the negative electrode, a negative charge is

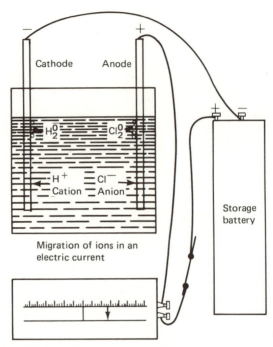

Fig. 4−3. Schematic outline of an experiment to determine ionic valence.

picked up by the ion, and a neutral hydrogen atom is restored, thus forming hydrogen gas on the negative electrode. A similar process takes place with chlorine: The chloride ion is attracted by the positive electrode, where, on contact, it loses an electron, thereby becoming electrically neutral and uniting with another chloride atom to form gaseous chlorine.

When oppositely charged ions form a molecule or a simple compound, such as HCl, it is the result of an *ionic valence*. The *valence number* for ions, found in tables, indicates the number of charges of an ion and is interpreted as the capacity for attracting opposite charges. Compounds formed by ionic valence are known as *polar compounds*.

Some atoms have more than one valence. Figure 4−2 shows the structure of atoms of chlorine and carbon and their valence.

Nonpolar compounds are shown, for example, with molecules of methane and carbon dioxide in Fig. 4−4. In these molecules, the bonding of atoms is not due to electrical forces. For methane, the electrons of hydrogen are attracted to the orbit of carbon, but not actually transferred to it. This is called sharing of the influence of electrons or *covalence*.

Through ionic valence or covalence the joining of atoms results in the formation of molecules. Molecules are pictured as the smallest particles or

Methane

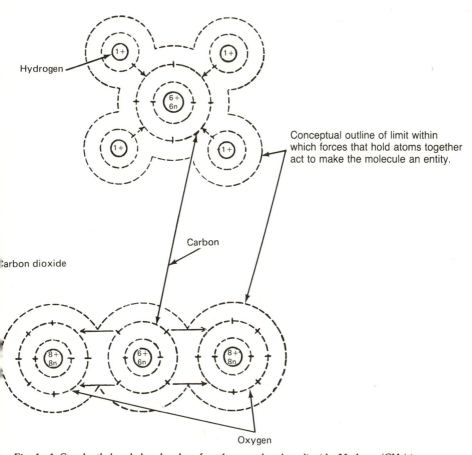

Hydrogen

Conceptual outline of limit within which forces that hold atoms together act to make the molecule an entity.

Carbon

Carbon dioxide

Oxygen

Fig. 4–4. Covalently bonded molecules of methane and carbon dioxide. Methane (CH_4) is a nonpolar compound. The electrons of the hydrogen atoms are shared with those of the carbon atom in the second orbit. Carbon dioxide (CO_2) is also a nonpolar compound. In both nonpolar compounds the sharing of influences of the electrons is taking place, and their atoms are joined by covalent bonds. The electrons in the second orbit of oxygen are shared with the two electrons of carbon.

ultimate physical units from which molecular chains are built up. There are other terms used in conjunction with molecular chains, such a simple compounds, mer, and monomer. A *mer* is the repeating structural unit in a high polymer. The *monomer* or single mer is a relatively simple compound that can react to form a polymer, which is composed of molecules containing many mers.

The "orbit" pictorial way of showing the joining or sharing of electrons and the forming of compounds is not practical as a normal procedure.

A method of representing the distribution of electrons in covalent molecules has been developed. A line between two atoms represents a pair of shared electrons and a dot represents an unshared electron. Two lines constitute a double bond of four shared electrons and three lines constitute a triple bond of six shared electrons.

It should be noted that molecules may also have ionic charges. It should also be indicated that molecules that are nonpolar may have unshared electrons, thus having the ability to combine further with other molecules. (See Fig. 4−5.)

Many of the simple compounds that are ready to combine or react with other atoms are called *unsaturated*, which means that they are ready to share electrons with other atoms or molecules.

Let us examine the simple compound of ethane (Fig. 4−6). In this structural formula we see the three hydrogen atoms and one of the carbon atoms combining to conform to the valence of four for carbon. This makes the ethane saturated.

If a hydrogen atom were left off each carbon atom, the valence of the carbon atoms would have to be satisfied by a double bond between the two carbon atoms. We now have the structural formula shown in Fig. 4−7. The compound is ethylene, which is the simplest unsaturated compound.

Acetylene, with a designation C_2H_2 (Fig. 4−5), has a triple bond between carbon atoms and two fewer hydrogen atoms. It thus has even greater unsaturation and is more reactive because of the greater number of unsatisfied valence bonds. Its structure is designated as $HC \equiv CH$.

Ethylene and acetylene belong to a group of compounds having a straight chain structure.

There is a large field of compounds in which the basic molecule contains a ring structure, usually of six carbon atoms. This is also known as the *cyclic* series and is often called the *aromatic group* (Fig. 4−8).

Benzene, C_6H_6, is the simplest member. Usually it is shown as a plain hexagon. In the majority cases, benzene behaves like a saturated compound. It is believed that the ring structure greatly reduces the reactivity of the unsaturated bonds below that anticipated from the reactions of such unsaturated compounds as ethylene and acetylene.

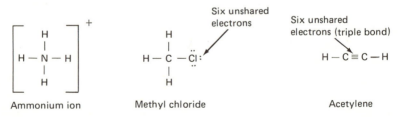

Fig. 4−5. **Polar charged molecules and nonpolar molecules with unshared electrons.**

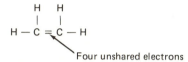

Fig. 4−6. **Ethane—an example of a saturated compound.**

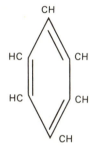

Four unshared electrons

Fig. 4−7. **Ethylene—the simplest unsaturated compound.**

CH

HC   CH

HC   CH

CH

Fig. 4−8. **Benzene—an example of the aromatic group.**

The simple compounds that we have analyzed are the starting materials for making many *polymers*. These polymers are compounds in which molecules of the same substance or other substances called monomers, have combined with each other in chain fashion to form a new compound of higher molecular weight and different physical properties. This process of molecular combination is called *polymerization*.

The structure of the polymer is generally described in terms of its structural unit, or simpler compounds. The structural units—ethylene, acetylene, benzene, etc.—having two or more bonding sites available, are linked through covalent bonds into a polymer molecule. This polymer can be of a linear structure, branched structure, planar network structure, or space network structure.

These polymer molecules vary in length and configuration depending on the polymerization method and reagents used in their formation. We can visualize them as wrapping strings with or without the short strings protruding from the body. These strings are intertwined and enmeshed with each other to form a finished polymer.

When the polymer contains a single repeating unit, such as polyethylene, it is a *homopolymer*, and when a polymer contains two units it is a *copolymer*, such as SAN. When it contains three structural units, it is called a *terpolymer*. ABS is sometimes designated as a terpolymer, although its structure is actually more complex.

The next step is to find out what different methods are employed to attain *polymerization* of a plastic material. On the basis of the type of chemical reactions that takes place, there are two classifications of polymerization, namely, *addition and condensation*.

1. Addition polymerization is obtained by the combination of basic substances (monomers) without loss. Addition polymers that are produced commercially are most always thermoplastic. The monomers have a double bond and with some exceptions (fluorinated compounds) always contain the methylene group, CH_2.

There are four basic methods of addition polymerization, each having certain advantages and disadvantages. The methods are listed below.

A. Bulk polymerization, in which one or more monomers are used, is the simplest of the four processes. A catalyst is also added to initiate polymerization. The mixture is usually stirred in a large reaction vessel with or without the addition of heat. The obvious advantage is purity of the polymer, as well as color and clarity, as, for example, in polystyrene. Economy, because only the raw material is involved, may also be a factor, but this may be outweighed by the fact that as the reaction proceeds stirring becomes more difficult because of the high viscosities. This requires expensive, heavy duty equipment. Most polymerization reactions are usually exothermic and, with large quantities of undiluted reactants, the problem of heat dispersion is acute. As the reaction proceeds, the viscosity increases and makes heat transfer more difficult.

B. Solution polymerization utilizes an inert solvent for the monomer and catalyst, and this solution is heated and stirred. In this method, the heat conductivity and stirring are eased, thereby making it adaptable to continuous processing. The cost of solvent and its recovery add to the expense of the process. The purity of polymer in this case is not as high as for bulk polymerization because, frequently, solvent fragments are found in the finished molecule. Traces of solvent are difficult and expensive to remove from the polymer. This process is advantageous for such purposes as coating or impregnating compositions where the polymer is used in solution form.

C. Emulsion polymerization usually employs water as the continuous medium with the monomer being dispersed in it. The catalysts, however, are soluble in the medium (water), and emulsifying agents such as sodium or potassium salts of fatty acids are added to stabilize the emulsion. After the reaction is completed the polymer is precipitated by coagulation of the emulsion and is washed to remove traces of the emulsifying agents. This is difficult to accomplish and some contaminants remain. Heat-transfer problems are minimal because the water is a good heat-transfer medium. This process produces the highest molecular weight polymers.

D. Suspension polymerization also employs water as the medium in which the monomer, as well as the monomer-soluble catalyst, are suspended. When the mixture is stirred, the monomer is broken into small droplets, each containing a little of the catalyst. A suspension stabilizer, such as small amounts of polyvinyl alcohol, is also included in the mixture in order to keep the droplets separate during reaction. After reaction, the stabilizers are removed by washing or flotation, leaving the polymer in the form of small beads. This method can be viewed as bulk polymerization on a miniature scale without the disadvantage of heat dispersion. The molecular weight of polymers in this method is also fairly high.

2. *Condensation polymerization* is a process where polymerization takes place through the interaction of one or more compounds; at the same time, the elimination of a simple compound occurs, such as water, ammonia, or hydrochloric acid. This method covers such polymers as polyamides, polyesters, alkyds, and phenolics. Condensation polymerization, therefore, can lead to either thermoplastic or thermosetting polymers. The important difference between condensation and addition polymerization is the formation of a by-product (e.g., water) that has to be removed during reaction, otherwise equilibrium will be reached and the reaction will stop, usually before a useful product is formed.

Copolymerization

Copolymerization can be considered an extension of addition polymerization and can take place with two or more different monomers. The resulting copolymer has different properties than those obtained from a mechanical mixture of the individual monomers. For example, when two different polymers are mechanically blended, most of the time the end result is a mixture whose properties are poor since the two polymers are incompatible. On the other hand, when we take two separate monomers and create conditions for their copolymerization, the result is a totally new material with distinctive properties.

There are many monomers that will not polymerize with themselves even under extreme conditions of temperature and pressure, but will copolymerize with others.

Copolymers may be

Random copolymers, in which different monomer units are randomly placed within the chain.

Alternating copolymers, in which polymers (say A and B) regularly alternate their position in the chain.

Block copolymers, in which relatively long blocks of different polymers are chemically bonded in the chain.

AAADDAADDDAAADD. . . .   Random copolymer

ADADADADADAD. . . .   Alternating copolymer

AAAAADDDDD. . . .   Block copolymer

DDDDDDDDDDDDDDDD. . . .   Graft copolymer
```
   A          A
   A          A
   A          A
   A          A
   A          A
   A          A
```

Fig. 4–9. Schematic representation of copolymers.

Graft copolymers, in which relatively short groups of one monomer attach themselves to a long backbone of another polymer. (See Fig. 4–9.)

CROSS-LINKING OR SIMILAR MODIFICATIONS

There are a number of polymers that can have their characteristics altered, although each one of them is composed of a specific formulation and each such formulation is reflected in certain defined properties. Some of the polymers can be modified not only by copolymerization, but also, for example, through permanent cross-linking achieved by chemicals or radiation as, for example, in cross-linked polyethylene for pipe and wire coating.

Alloying. Alloys are produced by melt mixing of two or more polymers or even of copolymers. The components are mutually soluble in one another and are compatible. Should compatibility be questionable, a compatibilizer is added to hold the materials together and to make them act as a homogeneous unit. Generally, alloys require stabilizers and plasticizers for processing. The following are some reasons that alloying is used: (1) to obtain a set of properties to meet a specific need; (2) to make these properties available at lowest cost; and (3) to improve processing. At present there are commercial alloys of ABS/polycarbonate; ABS/PVC; PVC/acrylic; polyphenylene oxide/polystyrene; ABS/polysulfone; SAN/polysulfone—and most likely there will be many more.

Polyblends form another source of plastic materials. These are polymers that are compatible and blended with each other. They are composed of a rigid phase into which another polymer is finely dispersed, i.e., polystyrene plus rubber. Commercially there are available rubber-modified polystyrenes, rubber-modified PVCs, and ABS.

Plastic materials, whether obtained by polymerization, copolymerization, cross-linking, alloying, or polyblending, are available for conversion into solid plastics (rigid or flexible), films, fibers, coatings, adhesives, and porous and nonporous materials.

COMPOSITION OF ALLIED POLYMERS

When polymerization takes place, the molecular chain becomes the basic constituent of the polymeric material. From the description of the various methods of polymerization, it becomes apparent that, with the numerous possible variables that may be encountered, it is reasonable to suspect that the lengths of the polymer chains may not be uniform and thus the molecular weights will vary. Therefore, when we speak of the molecular weight, we have in mind the average weight, which means considering the average chain length. The longer the molecular chain, the higher its weight; if we visualize chains miles long, intertwined with each other, we can conceive of the polymer that derives its characteristics from such a configuration as having many physical properties. We may conclude that *molecular weight* is an important factor in such properties. The applied polymer is not a collection of loosely placed molecular chains along side of each other, but has these chains bonded to each by secondary valence bonds, chemical bonds, and structural interaction bonds. The chemical bonds and valence bonds follow the same principles as in molecular formation. The structural interaction depends not only on shape of the molecule, i.e., whether it is branched or network shaped (see Fig. 4–10), but also on whether it is crystalline or amorphous (see Fig. 4–11).

In our discussion of molecules, we mentioned their threadlike configuration, neglecting the fact that under powerful magnification we could see that the basic molecules physically arrange themselves in the form of crystals. Many of

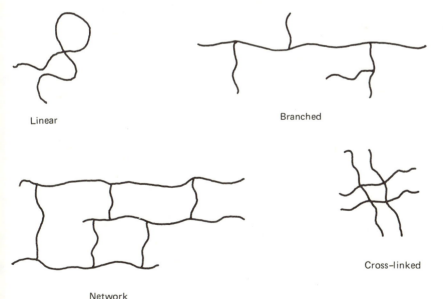

Linear

Branched

Network

Cross–linked

Fig. 4–10. Molecular shapes.

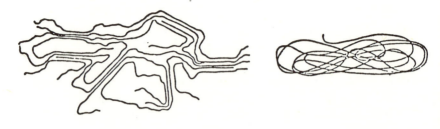

Crystalline   Amorphous

Fig. 4–11. Molecular structures.

the polymers have some degree of crystallinity, but none of them are 100% crystalline. The degree of crystallinity depends on how the chains are structurally constituted and ordered.

When the chains have a high degree of disorder, the polymer will tend to be amorphous, i.e., noncrystalline.

A crystalline polymer under the influence of an external stress will tend to orient itself so that its crystals are stretched in the direction of the force. The makeup of polymeric materials under powerful magnification could be compared to felts or fiberglass furnace filters as far as composition is concerned, except that molecular chains are thousands of times longer than the fibers in those two materials.

The processing conditions during polymerization are a determining factor for the size of the molecular chain. Since this is one of the major contributors to the physical properties of a polymer, chemists have been developing and applying several methods of establishing the weight of the molecular chains in the material offered for sale. They use this information as a measure of determining whether the process of polymerization was carried out along prescribed lines.

In conclusion it can be stated that the overall properties of a polymer depend on chemical composition and arrangement of atoms in the molecule; molecular weight or degree of polymerization, which means the number of molecules in the molecular chains; the nature and valence of bonds; the structural configuration of molecular chains; and the interaction between chains. The properties mostly affected by these factors are softening temperature and/or melting point; tensile strength and other mechanical properties; and melt viscosity.

Polymeric materials available for design consideration are classified into two groups: thermosetting and thermoplastic types.

1. Thermosetting types are where the resin undergoes a chemical reaction under the influence of heat, catalysts, ultraviolet light, etc., resulting in products that are relatively infusible in character, and are cross-linked materials that are not reversible, i.e., application of heat will not cause the material to soften (flow) and upon cooling it will not solidify. It is usually compared to the condition of an egg before and after it is hard boiled.

Typical members of this family are phenolics, epoxies, melamines, glyptals, and urea formaldehyde. Most of the thermoset materials are made with fillers, since the pure resins would be too brittle for practical applications. Some of the benefits derived from fillers are reduced shrinkage, reduced crazing, improved coefficient of expansion, and elimination of porosity. The selection of materials with appropriate fillers will also provide surface appearance, strength resistance to the environment, and favorable moldability.

The fillers frequently used in thermosets are wood flour, shell flour (nut shells), cotton flock, alpha-cellulose, chopped paper, macerated fabrics, synthetic fibers, carbon, asbestos, mica, quartz, silica, clay, glass fibers, and others that impart special characteristics to the molding material.

Thermosets can also incorporate graphite, Teflon, and silicone to reduce the coefficient of friction. It can be stated that fillers in thermosetting plastics are locked in within the cross-linked, large, and interconnected molecules that result in a permanent and stable material.

When comparing thermosetting and thermoplastic materials in a similar overall price class (material cost plus processing), the thermosets will deserve preferred standing when heat resistance, chemical resistance, rigidity, resistance to creep, hardness, flame resistance, and good electrical insulating characteristics are of primary concern. Any of these properties individually or in a group will deserve serious analysis during application considerations.

Fillers in thermosets are used in relatively high percentages (50% of each resin and filler), since they are the main contributors to the properties, while the resin acts as the cementing agent to hold the mixture together. The term "fillers" in this case also applies to the reinforcing agents, such as glass fibers and macerated fabrics.

The disadvantages of thermosetting materials in comparison with thermoplastics have been longer cycles; material wasted during an operation, including defective work, could not be recycled and thus was a total loss; in many materials colors were limited and lacking in stability; if toughness was needed, it normally led to higher costs; molds were subjected to higher wear, therefore, they required more frequent replacement; in general, one would require greater familiarity with the operations in order to be an effective producer; and, finally, most of the thermosets needed a postmolding deflashing operation.

With improvement in technology such as injection molding in conjunction with cold runner systems (runners are not wasted), the disadvantages are decreasing and with this decrease the competitive edge is more balanced.

2. Thermoplastic types are where the material has a long linear molecular structure and will repeatedly soften when heated, but will also become hard when cooled. Typical members of this family are polystyrene, polyethylene, polycarbonate, nylons, and vinyls. Each thermoplastic polymer has certain inherent characteristics which are due to the chemical composition. Commercial applications normally call for some features such as, for example, flammability, which may be lacking in the virgin polymer. In order to overcome some of the shortcomings of a thermoplastic, the industry has developed additives, fillers, extenders, and reinforcements that are compounded into the various polymers to generate the additional desired features.

Additives are incorporated not only for property modification, but also to improve processing. As a rule, additives do not appreciably impair the original properties of a polymer, although it is always prudent to verify this contention.

Some of the objectives of additives are to provide the following properties:

Flame retardance, which is intended to decrease the combustibility of materials by increasing the ignition temperature, reducing the rate of burning, minimizing the flame spread, and possibly reducing smoke evolution.

Ultraviolet light stability by the use of ultraviolet light absorbers, which prolong the useful life of products when exposed to UV radiation. Most of the thermoplastic materials are vulnerable to UV degradation, which causes breakdown of the polymer structure. Carbon black is one of the most effective UV absorbers and should be called for where color is not a problem.

Heat stability is gained by incorporating stabilizers that will prevent polymer degradation during elevated temperature processing, as well as during service life of a product when temperatures above normal are encountered.

Antioxidants provide protection against attack by oxygen if the chemical composition of the plastic is conducive to the degrading action of oxidation.

Other additives are available for such purposes as minimizing static electricity, protection against attack by microbial and fungicide agents, coloring agents, plasticizers, and lubricants for easy processing and other special purpose requirements.

Fillers and Reinforcements When thermoplastic materials are displaced by 10% or more by weight with an organic or inorganic material and the components are thoroughly mixed, a material with new properties is obtained. The bond or coupling between the components will determine the direction of the change in properties. With a good coupling agent or chemical coupling, we find improved tensile strength, higher modulus, equal or higher Izod impact, better creep resistance, higher resistance to heat distortion, lower shrinkage, and better processing capabilities. A poor bond will, during processing, tend to form resin-poor or resin-rich sections and, therefore, large variations in properties. Most other properties will not follow a favorable direction.

Extenders are introduced in plastic compounds mainly to substitute a low cost material for a part of the expensive polymer and thereby maintain a low selling price. In this category are mostly agricultural waste products, such as rice-hull fractions.

Fillers and reinforcements modify some properties, including resistance to creep, higher heat deflection temperature, lower elongation, higher specific gravity, higher modulus, and other characteristics depending on type and percentage of the filling agent.

Examples of filler materials are sand, quartz, novaculite, diatomaceous earth, talc, mica, asbestos, kaolin clay, wollastonite, and calcium carbonate.

Examples of reinforcements are fiberglass, asbestos fibers, carbon and graphite fibers, and boron fibers.

One important fact that is related to the overall benefit of fillers and reinforcements is the type of bond that exists between the polymer and its partial replacement. A favorable bond can be judged by comparing the Izod impact strength of the pure polymer against that of the composite. For example, the Celanese glass-coupled material GC25 has an Izod impact strength of 1.8, compared to the plain material M25 with an Izod impact strength of 1.6. Another case is of Hercules polypropylene filled with wollastonite that was surface modified with an effective coupling agent. The notched impact strength of the unfilled material was 0.35 compared to the filled polypropylene's notched impact strength of 0.43 or a 23% improvement.

The importance of a quality bond between the material components can be best appreciated when considering the finished product.

Some of the fillers and reinforcements may have an adverse influence on the coefficient of friction of the composite material, and, when it is necessary to overcome this deficiency, it is accomplished by the addition of a small percentage of Teflon, molybdenum disulfide, or similar agents.

The direction in which the properties that are influenced by fillers and reinforcement are as follows:

1. The modulus of elasticity or rigidity is higher.

2. The resistance to creep or deformation under load is higher.

3. The tensile strength is higher.

4. The heat conductivity is higher.

5. The resistance to elevated temperatures is higher.

6. The heat deflection temperature under load is higher.

7. The coefficient of thermal expansion is lower.

8. The specific gravity is higher.

9. The cost per cubic inch is higher.

10. The mold shrinkage is lower, which indicates a need for greater draft.

11. The impact strength at room or higher temperatures is usually lower. The reinforced materials are more brittle.

12. The weld line, whenever present, is less strong than parts without weld lines.

From a processing point of view, the filled and reinforced thermoplastics require certain considerations as follows:

1. The tooling will be subjected to a greater wear factor, therefore, for high activity parts the tool steel and its surface treatment will have to be chosen with better than normal abrasion resistance.

2. When the material is fed into cavities, the reinforcing agent will occupy a portion of the cross-sectional area of ducts, thereby necessitating an increase in gate, runner, and sprue in comparison with plain resin flow conditions.

3. Sharp corners and frequent bends in the flow path should be avoided. They may cause pockets of concentrated and segregated fiber areas that will cause stresses and uneven physical properties.

4. The injection pressures required for reinforced materials are usually higher, on the order of $10-30\%$ above the plain resin.

5. Cycles can be shorter in comparison with plain materials for two reasons: The reinforcing material does not have heat of fusion to be released, and, being a better heat conductor, it will bring about faster solidification.

6. The mold can run at higher temperatures because parts are more rigid at ejection.

7. On screw machines the speed of the screw and back pressure should be kept at the very minimum in order to avoid generating frictional heat and breaking up the lengths of fibers. Longer fibers will produce better properties.

Thus far we have become familiar with the general makeup of plastic materials without becoming involved in the specific characteristics and properties of each generic family of materials. This area will be covered in Chapter 5.

5. Polymers and Their Characteristics

INTRODUCTION

This chapter covers each polymeric material family. Its aim is to describe the material in a manner that will enable designers to select a suitable group for their contemplated purpose. This means that only information directly related to product design requirements will be included. Specifically, the description will cover outstanding use characteristics of commercially available polymers. It will indicate the grades within a material family, show data sheets of test properties as well as long-term information pertinent to product life. Furthermore, successful applications of end-use products will be pointed out, with the intention that such uses will suggest to the designer the appropriate interpretation of the above-mentioned test data. Finally, the processing methods that lend themselves to specific materials are also enumerated.

In discussing material characteristics, statements are made that "electrical properties are good" or "mechanical properties are on the high side," etc. These comments are intended to convey the thought that a material in its particular price class has been and is being selected for application due to the "good" or "favorable" property, in addition to other considerations. The descriptive contents should always be looked upon as an integral part of the data sheet, which is shown with each polymer family.

Figure 7−1 is a blank data sheet identical in all respects to those that appear in this chapter, except that the property headings have been correlated with ASTM test numbers. Figure 7−1, when reproduced, can be used by the designer to record data on present and future grades of material that are of interest to him or her but that are not listed in this volume. Such a procedure would enable the designer to compile an up-to-date file of information.

The ASTM designations are not repeated on the other data sheets in order to conserve space and to avoid unnecessary repetition.

The grades listed with their data sheets for each group of materials will be limited in number, since it would be impractical to list all of them in view of the continuous introduction of grades for special purpose needs.

The chemical-resistance phase of plastics is treated in a general way and should be adequate for design purposes. Once a material is decided upon for a product, the material supplier should be requested to check out conditions of usage, concentrations of chemicals, temperatures, and duration of exposure to determine if any ill effects may be expected on the performance or appearance of the designed item. It is always best to choose the specific grade of material and to keep in mind the extreme anticipated use conditions of a product when evaluating chemical resistance.

As a general rule, it is considered desirable to examine the properties of three or more materials before making a final choice. Material suppliers should be asked to participate in type and grade selection so that their experience is part of the input. The technology of manufacturing plastic materials has not advanced to a degree that the same polymer compounds supplied from various sources will deliver the same results in a product. As a matter of record, even each individual supplier furnishes his product under a batch number, so that any variation in conversion by the user can be tied down to the exact condition of the raw-material production. Taking into account manufacturing tolerances of the polymer, plus variables of equipment and procedure, it becomes apparent that checking several types of materials from the same and from different sources is an important part of material selection.

Experience has proven that the so-called interchangeable grades of materials have to be evaluated carefully by the designer as to their affect on the quality of a product. An important consideration as far as equivalent grade of material is concerned is its processing characteristics. There can be large differences in properties of a product and test data if the moldability features vary from grade to grade. It must always be remembered that test data have been obtained from simple and easy to mold shapes and do not necessarily reflect results in complex configurations.

The long-term behavior of a material, i.e., creep information, is given in the following form: Two or three values of apparent modulus are given after 100, 300, and 1000 hours of observation for the purpose of constructing a straight line on graph paper. The horizontal axis is "time" with a logarithmic scale and the vertical axis is the "apparent modulus" with a decimal inch scale or also a logarithmic scale. Such a straight line permits extension of the data for extrapolations into the needed time values. This method allows for conservation of space in this book and construction of the straight line on a scale that will be readily legible and suitable for the accurate reading of values. The reasoning behind this approach was based on the construction of over 600 curves with stress levels and temperatures of potential load applications for a great many grades of materials. It was observed that from 100 hours on all the curves closely follow a straight line. If information is needed for load duration of less than 100 hours, the apparent modulus as represented by the first point would provide an adequate safety factor and should be inserted in the appropriate formula. The details of applying creep data to design will be discussed in Chapter 7.

The problem of acquiring complete knowledge of candidate material grades should be resolved in cooperation with the raw material suppliers. It should be recognized that selection of the favorable materials is one of the three basic elements in a successful product—configuration design, material selection, and conversion into a finished product.

The order in which the descriptions of the materials is arranged is alphabetical according to the generic name. Wherever abbreviations of the generic names have been adopted, they are indicated in parentheses following the name, for example, polycarbonate (PC).

ABS (ACRYLONITRILE-BUTADIENE-STYRENE)

ABS, a thermoplastic and amorphous material, is a product of polymerizing three monomers. ABS is composed normally of two polymeric components that have been blended. These are: (1) styrene-acrylonitrile copolymer (SAN), which is the continuous phase, with (2) a dispersed phase of butadiene-containing rubber particles onto which SAN monomers are grafted. The graft component acts as a cementing agent since it is compatible with SAN continuous phase. Each monomer contributes to the properties of the finished material. Acrylonitrile contributes chemical resistance, heat resistance, and strength; butadiene assists in low-temperature property retention, impact resistance, and toughness; styrene adds to rigidity, gloss, and ease of processing.

By manipulating the percentage of each component, it is possible to attain materials with a broad range of properties. The variation occurs in such properties as specific gravity, impact resistance, heat deflection temperature, tensile—flexural strength, modulus of elasticity, flame retardance, plating, and transparency. With this multiplicity of variables available to the polymer chemist, different grades can be manufactured to meet special purpose requirements. These variables can be enhanced by additives and/or additional monomers, which will provide grades for high heat resistance.

Some of the important reasons for applying ABS are toughness, impact strength, good dimensional stability, rigidity, surface appearance, platability, colorability, and easy processing. The material is used for telephones; appliances; automobile interior and, when properly stabilized, exterior parts; sporting goods; luggage; and pipe and pipe fittings (drain, waste, and vent). ABS is resistant to many household and commercial chemicals. Processing is by injection, extrusion, blow molding, thermoforming, and other thermoplastic production methods.

Telephone set made of ABS.

ABS

PROPERTY	1. Cycolac DFA-R	2. Cycolac DH	3. Cycolac X37	4. Cycolac KJB	5. Cycoloy EHA	6. ABS/PVC alloy	7. AF-1004 (20% FG)	8. AF-1008 (40% FG)
PHYSICAL								
Mold shrinkage (in./in.)--in mils	6-8	5-7	5-7	6-8	6-8	3-5	1.5	1.0
Specific gravity	1.04	1.05	1.06	1.22	1.09	1.21	1.2	1.38
Specific volume (in.^{3}/lb)	26.6	26.4	26.1	22.7	25.4	23.2	23.1	20.1
Water absorption (%) 24 hr @ 73F (23C)							0.15	0.1
Water absorption (%) equilibrium, 73F (23C)							0.70	0.5
Haze (%)								
Transmittance (%)								
MECHANICAL								
Tensile strength (psi) yield	6,200	7,500	7,000	5,800	6,600	6,000	13,500	16,000
Tensile strength (psi) ultimate								
Elongation (%) yield	2.8						3-4	2-3
Elongation (%) rupture								
Tensile modulus (x10^5 psi)	3.4	3.8	3.5	3.2	3.5	3.3		
Flexural strength (psi)	11,400	13,000	12,000	10,000	11,400	10,200	17,500	20,000
Flexural modulus (x10^5 psi)	3.8	4.0	3.5	3.3	3.7	3.4	8.5	14.0
Compressive strength (psi)						7,400	13,500	16,000
Shear strength (psi)							7,000	8,000
Izod impact strength notched, 1/8 in.thick (ft-lb/in.)	4.0	4.4	3.0	4.0	6.9	12.5	1.5(1/4")	1.3(1/4")
Izod impact strength unnotched, 1/8 in.thick S type / L type							6.5(1/4")	5-6(1/4")
Tensile impact strength (ft-lb/in.^{2})								
Fatigue strength (psi @ cycles)								
Rockwell hardness	R108	R111	R109	R97	R111	R102	R124	R124
Deformation under load (%) @ 73F (23C) @ ___ F								
THERMAL								
Heat deflection temperature (F) @ 66 psi	201	215	250	200	247	166	225	235
Heat deflection temperature (F) @ 264 psi	184	201	230	190	232	150	215	225
Specific heat (Btu/lb/F or Btu/lb/C)								
Thermal conductivity (Btu/hr/ft^2/F/in.)								
Coeff. therm. exp.(x10^{-5})(in./in./F or /C)	4.9	3.9	4.4		3.5	8.2	2	1.2
Brittleness temperature (F)								
Flammability UL Standard 94	94HB	94HB	94HB	94V-0			94HB	94HB
Oxygen index								
ELECTRICAL								
Dielectric strength (V/mil)	821	377	937	400				
Dielectric constant 60H / 10^{6}H								
Power factor 60H / 10^{6}H								
Volume resistivity (ohm-cm)	2.9x10^9	9.3x10^7	7x10^9	1.4x10^8				

ABS (continued)

PROPERTY	stran	ustran R	ustran	ustran 08	G 298-ed	adon 112	adon 127	adon 160
PHYSICAL								
Mold shrinkage (in./in.)--in mils	4-6	4-6		4-6		5	5	5
Specific gravity	1.04	1.05	1.2	1.06		1.07	1.07	1.07
Specific volume (in.³/lb)	26.6	26.4	23.1	26.1		25.9	25.9	25.9
Water absorption (%) 24 hr @ 73F (23C)						0.5	0.5	0.5
equilibrium, 73F (23C)			0.09			0.3	0.3	0.3
Haze (%)								
Transmittance (%)								
MECHANICAL								
Tensile strength (psi) yield	5,000	6,800	6,600	6,950		5,200	5,000	6,000
ultimate				6,100	9,000			
Elongation (%) yield				2.5		1.8	1.7	1.8
rupture				18		40	40	30
Tensile modulus (x10⁵ psi)	2.6	3.6	3.9	4.15	11.0	3.0	3.1	3.5
Flexural strength (psi)	7,300	11,000	11,000	12,500	17,000	8,000	8,400	9,000
Flexural modulus (x10⁵ psi)	2.8	3.7	3.8	4.25	18.2	3.2	3.4	3.2
Compressive strength (psi)								
Shear strength (psi)								
Izod impact strength notched, 1/8 in.thick (ft-lb/in.)	7.0	5.0	5.2	5.2	6.2	6.0	5.5	5.0
unnotched, 1/8 in.thick (ft-lb/in.)								
Tensile impact strength (ft-lb/in.²) S type								
L type								
Fatigue strength (psi @ cycles)								
Rockwell hardness	R105	R109	R110	R111		R109	R109	R109
Deformation under load (%) @ 73F (23C) @ 66 psi								
@ 264 psi								
THERMAL								
Heat deflection temperature (F) @ 66 psi	191	216	176	217	290	223	235	216
@ 264 psi	180	208	162	209				
Specific heat (Btu/lb/F or Btu/lb/C)								
Thermal conductivity (Btu/hr/ft²/F/in.)						0.0066	0.0063	0.0066
Coeff. therm. exp(x10⁻⁵) (in./in./F or /C)	5.3	4.7	4.7	4.5		4.7	5.2	4.7
Brittleness temperature (F)								
Flammability UL Standard 94			94V-0			94HB	94HB	94HB
Oxygen index			28					
ELECTRICAL								
Dielectric strength (V/mil) 60H						430	420	430
10⁶H								
Dielectric constant 60H						3.10	3.15	3.10
10⁶H						3.08	3.03	3.07
Power factor 60H / 10⁶H								
Volume resistivity (ohm-cm)						10¹⁵	10¹⁵	10¹⁵
Arc resistance (sec), tungsten electrodes								

ABS

Items 1–6 are "Cycolac" brand grades, trademarks of Borg Warner Chemicals, Inc. *Items 7 and 8* are glass-reinforced materials of LNP Corp. *Items 9–13* are "Lustran" grades and *14–16* are "Cadon" grades, both trademarks of Monsanto Co.

Item 1 can be used for housewares, appliances, hardware, and office–home furnishings.

Item 2 can be used where high heat, rigidity, and tensile and medium impact strengths are factors.

Item 3 can be used where heat resistances of 20°F (11°C) above most ABS grades and good impact strength are the main requirements.

Item 4 can be used for electrical uses requiring regulatory agency approval (business machines, electrical appliances, power tools, etc.) as well as for UV resistance to color change in indoor applications.

Item 5 can be used where impact strength, superior thermal cycle performance, and resistance to warpage are considerations.

Item 6 can be used for conditions which involve flame retardance, impact strength, and chemical resistance not found in other grades.

Items 7 and 8 are glass reinforced and can be used where high modulus and resistance to creep are the predominant requirements.

Item 9 has good impact strength, resilience, and good moldability and can be used for lawn and garden equipment, power tool housings, and appliances.

Item 10 can be used where high heat resistance and toughness are needed, such as for seatbelt buckles, light housings, and vacuum cleaner components.

Item 11 can be used where flame retardance is a major consideration.

Items 12 and 13 are plating grades, showing improvements in some properties after plating. The plating thickness is 0.0015"; copper, 0.0008"; nickel, 0.0007"; and chrome, 0.00001". The properties would change with the change in the metal thicknesses of plating.

Items 14–16 are relatively new additions to the family and are impact-modified styrene maleic anhydride terpolymers, with a good balance of heat resistance and impact strength. They fill the gap between regular ABS grades and higher-performance materials. They represent high impact strength, medium high heat resistance; medium high impact strength, high heat resistance; and general purpose medium high heat resistance grades, respectively.

The companies listed as well as other suppliers make additional grades which can be found in their commercial literature. The data and applications were also adapted from these same publications. Additional information can also be found in the Monsanto Co.'s *Lustran Engineering Design Data.* The creep data were obtained from a test in flexure with the specimen supported at 2-in. spacing.

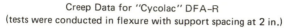

Creep Data for "Cycolac" DFA–R
(tests were conducted in flexure with support spacing at 2 in.)

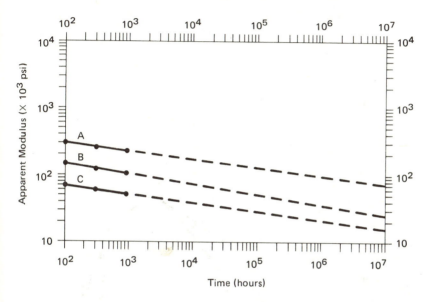

A, 73° F (23° C) and 1000 psi; B, 120° F (49° C) and 1000 psi; C, 140° F (60° C) and 1000 psi. Dotted lines indicate extrapolated values.

ABS-PC Alloy Resin (ABS–Polycarbonate Alloy)

This combination alloy displays the best properties of each component. It has good heat resistance, has outstanding impact strength, and is easy to process, as well as having good dimensional stability. Its applications extend to products where ABS's heat and impact characteristics are inadequate for the use conditions.

ABS-PVC Alloy Resin (ABS–Polyvinyl Chloride Alloy)

This alloy displays good impact strength, chemical resistance, toughness, flame retardance for applications where ABS alone is lacking in the above characteristics.

ACETAL COPOLYMER

Acetal copolymer is a thermoplastic, crystalline material. Chemically, the resin is a copolymer of trioxane with small amounts of a comonomer that randomly distributes carbon–carbon bonds in the polymer chain. Acetal copolymer has good rigidity, high strength toughness and resistance to creep and fatigue along with an inherently low coefficient of friction and high abrasion resistance; its dimensional stability is excellent. Electrically the material has a low dielectric constant and dissipation factor that are constant over the range of use temperatures. The chemical resistance is excellent against common solvents, lubricants or gasoline, acid, and alkalis. Its good properties are retained in hot air up to 220°F (104°C) and in hot water up to 180°F (82°C) for prolonged periods of time; higher temperatures can be tolerated for intermittent use.

Special grades are available for contact with foods. Another grade is formulated for exceptionally low friction characteristics for such parts as bearings and moving parts. Glass-reinforced grades have a unique coupling property of polymer and glass fiber and are observed to show an increasing Izod impact over the plain polymer.

Acetal copolymer is translucent white and can be colored in a number of standard shades. Applications are for plumbing valves, pumps, faucets, small gears, bearings, electrical switching devices, seat-belt hardware, pens and pencils, aerosol bottles, cams for mechanical devices, etc. The favorable flexural properties permit the application of the material for spring-type actions.

Processing can be performed by injection, extrusion, blow molding, and rotomolding.

The acetal materials (copolymer and homopolymer) are similar in performance and are frequently used as interchangeable grades. In some critical applications, it is possible to find enough difference to favor one or the other material.

Typical parts made from acetal copolymer.

ACETAL COPOLYMER

Category	Property	Celcon 5-04	Celcon |-04	Celcon)-08	Celcon 70-04	Celcon -25A	KFX1002 (% GF)	KFX1006 (% FG)	KFX1008 (% MG)
PHYSICAL	Mold shrinkage (in./in.)--in mils	18-22	18-22	18-22	18-22	4-18	8-10	3-16	6-8
	Specific gravity	1.41	1.41	1.41	1.41	1.59	1.47	1.63	1.71
	Specific volume (in.^{3}/lb)	19.7	19.7	19.7	19.7	17.54	18.9	17.0	16.2
	Water absorption (%) 24 hr @ 73F (23C)	0.22	0.22	0.22	0.22	0.29	0.24	0.30	0.30
	Water absorption (%) equilibrium, 73F (23C)	0.80	0.80	0.80	0.80				
	Haze (%)								
	Transmittance (%)								
MECHANICAL	Tensile strength (psi) yield	8,800	8,800	8,800	8,800	16,000	12,500	19,500	13,500
	Tensile strength (psi) ultimate								
	Elongation (%) yield					2-3			
	Elongation (%) rupture	>250	>250	>250	>250	12	6	3-4	3
	Tensile modulus (x10^5 psi)	4.1	4.1	4.1	4.1	10.5			
	Flexural strength (psi)	13,000	13,000	13,000	13,000	13,000	18,500	29,000	21,000
	Flexural modulus (x10^5 psi)	3.75	3.75	3.75	3.75		5.5	14.0	11.0
	Compressive strength (psi) (10% D)	16,000	16,000	16,000	16,000				
	Shear strength (psi)	7,700	7,700	7,700	7,700		8,000	9,000	8,200
	Izod impact strength notched, 1/8 in. thick (ft-lb/in.)	1.5	1.3	1.3	1.1	1.1	1.4	1.8	1.1
	Izod impact strength unnotched, 1/8 in. thick (ft-lb/in.)						10.0	8-10	6-7
	Tensile impact strength (ft-lb/in.^{2}) S type	90	70	70	60	50			
	Tensile impact strength (ft-lb/in.^{2}) L type								
	Fatigue strength (psi @ cycles)								
	Rockwell hardness	M80	M80	M80	M80		M82	M86	M84
	Deformation under load (%) @ 73F (23C) @ 2000 @ 122F	1.0	1.0	1.0	1.0	0.6	0.85	0.6	0.8
THERMAL	Heat deflection temperature (F) @ 66 psi	316							
	Heat deflection temperature (F) @ 264 psi	230				322	300	330	310
	Specific heat (Btu/lb/F or Btu/lb/C)	0.35	0.35	0.35	0.35				
	Thermal conductivity (Btu/hr/ft^2/F/in.)	1.6	1.6	1.6	1.6				
	Coeff. therm. exp.(x10^{-5})(in./in./F or /C)	4.7	4.7	4.7	4.7		2.6	2.2	2.4
	Brittleness temperature (F)								
	Flammability UL Standard 94						94HB	94HB	94HB
	Flammability Oxygen index								
ELECTRICAL	Dielectric strength (V/mil)	500	500	500	500	600			
	Dielectric constant 60H	3.7	3.7	3.7	3.7	4.04			
	Dielectric constant 10^{6}H								
	Power factor 60H	0.006	0.006	0.006	0.006	0.007			
	Power factor 10^{6}H								
	Volume resistivity (ohm-cm)	10^{14}	10^{14}	10^{14}	10^{14}	0.8x10^{14}			
	Arc resistance (sec), tungsten electrodes	240	240	240	240	142			

Acetal Copolymer

Items 1−5 are "Celcon" grades, trademarks of Celanese Plastics Co. *Items 6−8* are products of LNP Corp.

Item 1 can be used for injection-molded parts requiring high toughness and elongation.

Item 2 is a general purpose injection-molding grade.

Item 3 is a UV-stabilized grade for outdoor exposure.

Item 4 is internally lubricated for ease of filling multicavity molds.

Item 5 is a 25% glass-filled grade with an outstanding coupling agent that results in superior stiffness, tensile strength, and creep resistance. It can be used for windshield wiper blades and for other components where its high stiffness and strength are required.

Item 6−8 are glass-reinforced grades with a chemical coupling agent that significantly improves its physical properties in comparison with standard reinforced grades. In addition, *Item 8* is filled with a mixture of glass particles that results in a minimal difference in shrinkage when measured in the direction of and perpendicular to the flow.

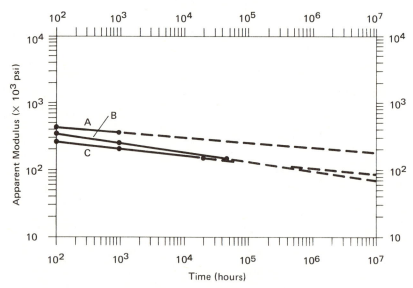

Creep Data for Celcon
(tests were conducted in flexure with support spacing at 2 in.)

A, GC25A at 180°F (82°C) and 500 psi; B, M90 at 73°F (23°C) and 500 psi; C, M90 at 73°F (23°C) and 5000 psi. Dotted lines indicate extrapolated values.

The following illustrations show relevant test data of the material. All data and illustrations have been adapted from the suppliers' commercial literature. Additional information can be found in the Celanese Co.'s *Design Manual for Celcon Acetal Copolymer*. The creep data were obtained from a test in flexure with the specimen supported at 2-in. spacing.

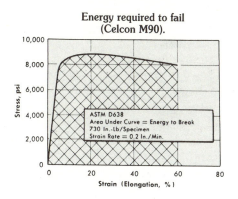

Energy required to fail (Celcon M90).

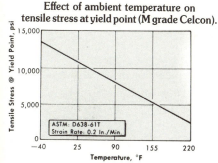

Effect of ambient temperature on tensile stress at yield point (M grade Celcon).

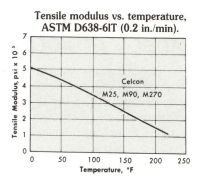

Tensile modulus vs. temperature, ASTM D638-6IT (0.2 in./min).

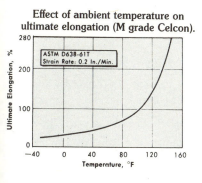

Effect of ambient temperature on ultimate elongation (M grade Celcon).

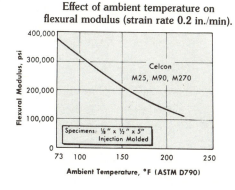

Effect of ambient temperature on flexural modulus (strain rate 0.2 in./min).

Celcon M90 stress–strain curve
at 73°F (23°C).

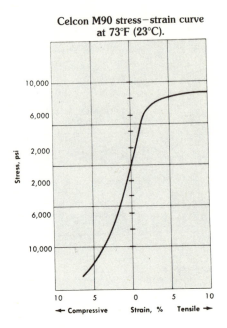

ACETAL HOMOPOLYMER

Acetal homopolymer is a thermoplastic crystalline material. Chemically, the resin is a high-molecular-weight, stable, linear polymer of formaldehyde. It has good rigidity, strength, toughness, and resistance to creep and fatigue. The dimensional stability is also favorable. Moisture has little effect on the mechanical properties or on the dimensions. The low coefficient of friction and resistance to abrasion are major factors in applications of the material. Electrically, the resin has a low dissipation factor and dielectric constant; these conditions prevail up to 250°F (121°C). The material is resistant to organic compounds and common solvents, but is not recommended for extended contact with strong acids or bases.

There are three basic grades available, which are differentiated by molecular weight. The subdivisions of the main grades are:

1. High productivity resins.

2. An "AF" resin, which contains high strength fibers of Teflon for improved frictional and wear properties in bearings and moving parts.

3. An improved friction and wear-resistant grade through the use of a chemical lubricant.

4. A glass-reinforced grade for superior mechanical properties.

Applications in the automotive field are for components in the fuel system, seat belts, steering columns, and window activating mechanisms. In the plumbing field the material is used for shower heads, mixing valves, flushometer parts, and faucets. Lawn sprinklers, garden sprayers, stereo and video tape cassette components, zippers for clothing, and a variety of machinery components also take advantage of the desirable material properties.

Processing is carried out by injection, extrusion, blow molding, and roto-molding.

"Delrin" clock frame incorporates many features which in metal require additional parts. Wheels and gears are also molded of "Delrin."

Creep Data for Delrin
(tests were conducted in flexure with support spacing at 4 in.)

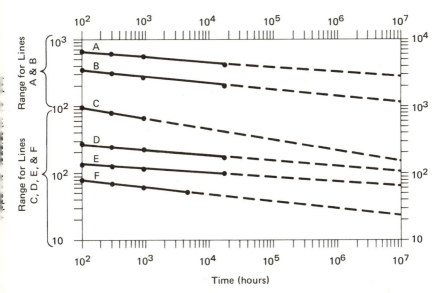

Time (hours)

A, Delrin 570 at 73°F (23°C) and 1000 psi; B, Delrin 570 at 140°F (60°C) and 1000 psi; C, Delrin 570 at 195°F (91°C) and 2000 psi; D, Delrin 100, 500, and 900 at 73°F (23°C) and 1000 psi; E, Delrin 100, 500, and 900 at 140°F (60°C) and 1000 psi; F, Delrin 100, 500, and 900 at 212°F (100°C) and 1000 psi. Dotted lines indicate extrapolated values.

ACETAL HOMOPOLYMER

	PROPERTY	1. Delrin 100	2. Delrin 500	3. Delrin 900	4. Delrin 570 (20% GR)	5. Delrin 500Cl (chemically lubricated)	6. Delrin 100AF (with Teflon fibers)	7. Delrin 500AF (with Teflon fibers)
PHYSICAL	Mold shrinkage (in./in.)--in mils	15-35 (see note 1)						
	Specific gravity	1.42	1.42	1.42	1.56	1.42	1.54	1.54
	Specific volume (in.³/lb)	19.5	19.5	19.5	17.76	19.5	18	18
	Water absorption (%) 24 hr @ 73F (23C)	0.25	0.25	0.25	0.25	0.27	0.20	0.20
	Water absorption (%) equilibrium, 73F (23C)	0.9	0.9	0.9	1.0	1.0	0.72	0.72
	Haze (%)							
	Transmittance (%)							
MECHANICAL	Tensile strength (psi) yield / ultimate	10,000	10,000	10,000	8,500	9,500	7,600	6,900
	Elongation (%) yield / rupture	75	4.0	2.5	12	40	22	15
	Tensile modulus (x10⁵ psi)	4.5	4.5	4.5	9.0	4.5	4.2	4.2
	Flexural strength (psi)	14,300	14,100	14,000	10,700	13,000	10,500	10,300
	Flexural modulus (x10⁵ psi)	3.8	4.1	4.3	7.3	4.0	3.4	3.5
	Compressive strength (psi) @10% deflection	18,000	18,000	17,600	18,000	15,500	13,000	13,000
	Shear strength (psi)	9,500	9,500	9,500	9,500	9,500	8,000	8,000
	Izod impact strength notched,1/8 in.thick (ft-lb/in.²)	2.3	1.4	1.3	0.8	1.4	1.2	0.7
	Izod impact strength unnotched,1/8 in.thick	>100	24	16				
	Tensile impact strength (ft-lb/in.²) S type / L type	170	100	70	33	100	50	32
	Fatigue strength (psi @ 10⁶ cycles)	4,700	4,500	4,600	4,500	4,000	3,600	3,500
	Rockwell hardness	R120	R120	R120	R118	R120	R118	R118
	Deformation under load (%) @ 2000@122 F @ 73F (23C)	0.5	0.5	0.5	0.4	0.7	0.6	0.6
THERMAL	Heat deflection temperature (F) @ 66 psi	342	342	342	345	338	334	334
	Heat deflection temperature (F) @ 264 psi	277	277	277	316	255	244	244
	Specific heat (Btu/lb/F or Btu/lb/C)	0.35	0.35	0.35		0.35		
	Thermal conductivity (Btu/hr/ft²/F/in.)	2.6	2.6	2.6				
	Coeff. therm. exp.(x10⁻⁵)(in./in./F or /C)	5.8	5.8	5.8	3.2	5.8	5.8	5.8
	Brittleness temperature (F)							
	Flammability UL Standard 94 / Oxygen index	94HB	94HB	94HB	94HB	94HB	94HB	94HB
ELECTRICAL	Dielectric strength (V/mil)	500	500	500	490	400	400	400
	Dielectric constant 60H / 10⁶H	3.7	3.7	3.7	3.9	3.5	3.1	3.1
	Power factor 60H / 10⁶H	0.005	0.005	0.005	0.005	0.006	0.009	0.009
	Volume resistivity (ohm-cm)	10¹⁵	10¹⁵	10¹⁵	5x10¹⁴	5x10¹⁴	3x10¹⁶	3x10¹⁶

Acetal Homopolymer

Delrin is a trademark of the Du Pont de Nemours & Co.

Item 1 is the toughest.

Item 2 is a general purpose grade.

Item 3 has high flow characteristics.

Item 4 is a 20% glass-reinforced grade.

Item 5 is chemically lubricated.

Items 6 and 7 are impregnated with Teflon fibers.

Notes

1. The shrinkage rate indicated is too broad for practical consideration. The average of the two values is a fair general figure for most preliminary estimates. However, accurate data can be obtained from the *Du Pont Molding Manual for Delrin*, where a nomograph shows the relationship among part thickness, gate area, mold temperature, injection pressure, and corresponding shrinkage.
2. The coefficient of friction was determined by a thrust washer test at 10 feet per minute and 300 psi rubbing pressure against AISI 1080 carbon steel, Rc 20, finished to 16 microinches. The test was conducted on a rotating disc tester. The results were 0.35 for *Items 1 – 4*, 0.20 for *Item 5*, and 0.14 for *Items 6 and 7*.
3. The creep data shown were obtained from a test in flexure with the specimen support at 4-in. spacing.
4. The following illustrations show relevant test data. All the data and illustrations have been adapted from the *Design Handbook of Du Pont Delrin*, where considerably more data, encompassing various applicational considerations, can be found.

Long-term behavior of "Delrin" under load at 73°F (23°C), air.

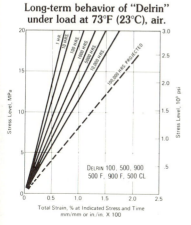

Long-term behavior of "Delrin" under load at 113°F (45°C), air.

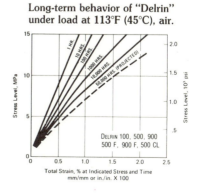

Long-term behavior of "Delrin" under load at 185°F (85°C), air.

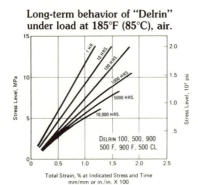

Total Strain, % at Indicated Stress and Time
mm/mm or in./in. X 100

Long-term behavior of "Delrin" under load at 212°F (100°C), air.

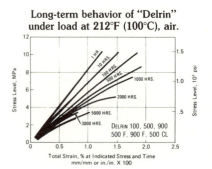

Total Strain, % at Indicated Stress and Time
mm/mm or in./in. X 100

Tensile secant modulus for "Delrin" vs temperature and cross-head speed (at 1% strain).

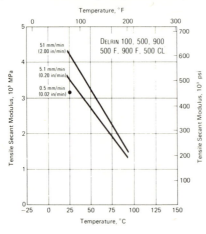

Tensile yield stength of "Delrin" vs temperature cross-head speed of 0.2 in./min.

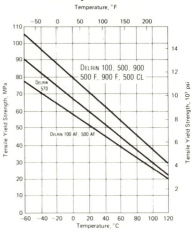

Stress—Strain curves for "Delrin" in tension and compression at 73°F (23°C).

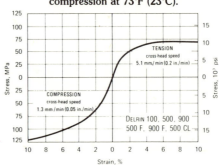

Strain, %

Stress–strain curves for "Delrin" acetal resins at various temperatures and rates of loading (ASTM D638).

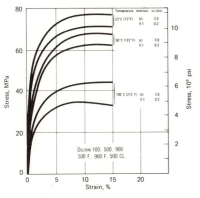

Stress vs strain for "Delrin" cross-head speed of (0.2 in./min).

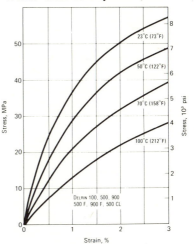

Tensile stress relaxation of "Delrin" under constant strain at 73°F (23°C).

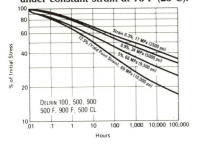

Flexural modulus of "Delrin" vs temperature at cross-head speed of 0.05 in./min.

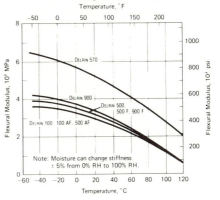

ACRYLIC

Acrylic resin is a thermoplastic, amorphous material. Chemically, the resin is made by the polymerization of acrylic ester monomers, the most common one being methyl methacrylate.

Its outstanding characteristic is crystal clarity comparable in this respect to glass and resistance to degradation in outdoor environments, as proven over many years of favorable service performance.

Exposure to fluorescent, incandescent, or mercury vapor radiation produces no ill effects on the material. Rigidity, intermediate toughness, and relatively good mar resistance are additional properties that bring about many applications. Color possibilities are unlimited. Service temperatures extend over a broad range, i.e., $-40-240°F$ ($-40-116°C$).

Acrylic is not affected by most detergents and cleaning agents nor by solutions of alkalis, acids, or aliphatic hydrocarbons. Exposure to ketones, esters, and chlorinated and aromatic hydrocarbons should be avoided.

The resin is combustible, and this shortcoming should be considered when contemplating applications. When subjected to a continuous load, the stress level should not exceed 1000 psi since under these circumstances surface crazing may be initiated. Intermittent loads at above or below stress level will not have an adverse effect.

Acrylics are processed by casting or by extruding sheets for fabrication into required shapes and also by injection molding and bar extrusion of formed shapes. Sheet materials may also have a silicate coating applied that appreciably improves the scratch resistance of the surface.

The Izod impact resistance of the unmodified resin is low in comparison with most thermoplastics. Alloying with rubber will improve the Izod impact strength up to about 20 times, and, when the alloy consists of acrylic and polyvinyl chloride, flame retardance is added. Alloyed sheet materials have excellent thermoforming properties and contribute to a large extension of uses; however, they are opaque.

Acrylics are used for indoor and outdoor signs of every type; for glazing against vandalism; for automotive parts such as tail lights, instrument panels, name-plates, dials; and for innumerable other products.

The alloyed sheets of proper grade are thermoformed for packaging of food, medical, and pharmaceutical products.

ACRYLIC

PROPERTY		ydex 100	lexiglas	Lexiglas	Lexiglas	Lexiglas	cite 148	cite 147	cite 140	cite 129
PHYSICAL	Mold shrinkage (in./in.)--in mils		2-8	2-6	2-6	2-8	3-7	3-7	3-7	2-7
	Specific gravity	1.35	1.15	1.18	1.18	1.17	1.19	1.19	1.18	1.18
	Specific volume (in.^{3}/lb)	20.5	24.0	23.5	23.5	23.7	23.3	23.3	23.5	23.5
	Water absorption (%) 24 hr @ 73F (23C)	0.06	0.3	0.3	0.3	0.3	0.3	0.3	0.3	0.3
	Water absorption equilibrium, 73F (23C)									
	Haze (%)					3.3	<3	<3	<3	<3
	Transmittance (%)		90	92	92	92	92	92	92	92
MECHANICAL	Tensile strength (psi) yield		5,500							
	Tensile strength ultimate	6,000	5,400	8,700	9,600	7,000	11,000	10,500	10,500	10,000
	Elongation (%) yield		5							
	Elongation rupture	>100	35				5	4-7	3-5	3-5
	Tensile modulus (x10^5 psi)	3.1	2.2	3.8	4.1	3.25	4.5	4.5	4.5	4.5
	Flexural strength (psi)	9,700	7,000	14,000	15,000	10,500	17,000	17,000	16,000	15,000
	Flexural modulus (x10^5 psi)	3.3				3.25	4.5	4.5	4.5	4.5
	Compressive strength (psi)	8,000	6,000	14,000	14,500	10,500				
	Shear strength (psi)	4,500	6,470							
	Izod impact strength notched,1/4 in. thick (ft-lb/in.)	10-15	1.2	0.4	0.4	0.6	0.43	0.4	0.3	0.3
	Izod unnotched,1/8 in. thick									
	Tensile impact strength (ft-lb/in.^{2}) S type	75								
	L type									
	Fatigue strength (psi @ ___ cycles)									
	Rockwell hardness	R90	R99			M68	M95	M95	M95	M90
	Deformation under load (%) @ 73F (23C) @ 66 psi / @2000@ 122 F @ 264 psi						0.6	0.3	0.35	0.7
THERMAL	Heat deflection temperature (F)	177/165	180/170	180/160	180/170	190	210/198	210/198	210/198	190/180
	Specific heat (Btu/lb/F or Btu/lb/C)	0.29					0.35	0.35	0.35	0.35
	Thermal conductivity (Btu/hr/ft^2/F/in.)	1.1	1.5				1.4	1.4	1.4	1.4
	Coeff. therm. exp.(x10^{-5})(in./in./F or /C)	4.3	5.6				3	3	3	4
	Brittleness temperature (F)									
	Flammability UL Standard 94	94V-0	94HB	94HB	94HB	94HB				
	Oxygen index	40								
ELECTRICAL	Dielectric strength (V/mil)	>430	383	500	500		400	400	400	400
	Dielectric constant 60H	3.86	3.9	3.8	3.7		3.3	3.3	3.5	3.9
	Dielectric constant 10^{6}H	3.44					2.7	2.7	2.7	2.9
	Power factor 60H	0.02	0.04	0.04	0.04		0.06	0.06	0.06	0.04
	Power factor 10^{6}H	0.027					0.02	0.02	0.02	0.03
	Volume resistivity (ohm-cm)						10^{15}	10^{15}	10^{15}	10^{14}
	Arc resistance (sec), tungsten electrodes	80					NO TRACKING			

Acrylic

Items 1–4 are "Lucite" materials, trademarks of Du Pont. *Items 5–8*, "Plexiglas" grades, and *Item 9*, an alloy of acrylic and PVC sheet known as "Kydex," are tradenames of Rohm and Haas Co.

Item 1 is a medium flow material suitable for flashlight lenses, lighting louvers, and instrument cases where maximum heat resistance is not required.

Item 2 has good heat resistance and is used for automotive parts such as horn buttons, hood ornaments, taillamp lenses, and instrument panel dials.

Items 3 and 4 possess higher toughness than the preceding grades and are used for sun lenses in eye glasses and TV screen shields.

Item 5 is a tough acrylic grade suitable for medical and leisure products.

Items 6 and 7 are noted for resistance to degradation when exposed to sustained UV light.

Item 8 has a high impact strength, up to 10 times higher than conventional grades of acrylics, without sacrificing other desirable characteristics.

Item 9 is an alloy sheet material used for thermoforming of such products as machine and equipment housings, trays, tote boxes, and safety helmets. The acrylic component in the alloy contributes to superior rigidity and formability, while PVC adds to the toughness, chemical resistance, and good surface finish.

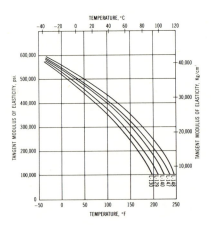

Tangent modulus of elasticity of Lucite vs temperature (ASTM D638).

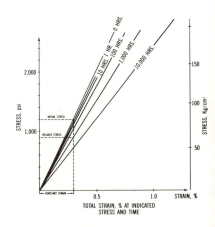

Long-term behavior of Lucite 140 under load at 77°F (25°C), air.

Additional grades are available from the above and other suppliers.

The following illustrations point to additional data pertaining to the behavior of the material. The data and illustrations have been adapted from suppliers' literature. More information of a detailed nature can be found in DuPont's *Lucite Design Handbook*.

Long-term behavior of Lucite 140 under load at 100°F (38°C), air.

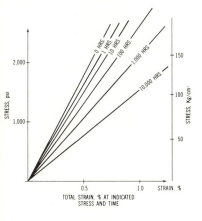

Long-term behavior of Lucite 140 under load at 140°F(60°C), air.

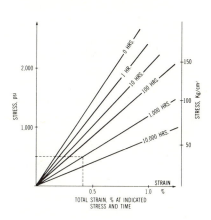

Stress–strain curves for Lucite 147 at two temperatures

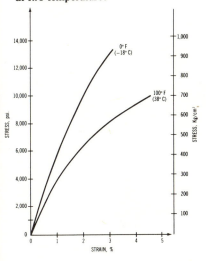

Tensile strength of Lucite vs temperature (ASTM D638).

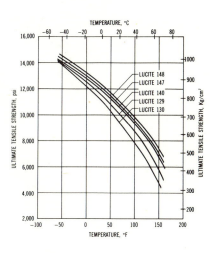

ALLYL – DIALLYL PHTHALATE (ALLYL-DAP)

Chemically, this resin consists of monomers and prepolymers of diallyl phthalate and diallyl isophthalate. The material is commercially available as a diallyl phthalate compound that is thermosetting and possesses electrical properties needed in critical military applications such as connectors, terminal blocks for communications, computers, and aerospace systems. The molding grades are available in mineral-, glass-, and synthetic-fiber-filled grades.

These compounds have good electrical properties under conditions of high humidity and high temperatures; they also have stable low loss factors, high surface and volume resistivity, good arc and track resistance, and good dielectric strength values up to 374°F (190°C). The material also exhibits the following properties: good dimensional stability coupled with good mechanical strength, heat resistance, favorable flammability rating, and resistance to chemicals.

Allylic resins are also used for impregnating laminated papers and fabrics prior to molding for decorating as well as for functional electrical property purposes.

A material in this group, known as CR-39 made by PPG Industries, Inc., has outstanding scratch and mar resistance along with optical clarity, toughness, and light weight. It is used extensively for eye lenses where weight is a problem and/or a danger of lens breakage is of concern. It is a cast material in sheet form and lends itself to fabrication for lens prescriptions. The scratch and mar resistance of CR-39 has been used as a standard of comparison of other plastic materials and coatings with regard to these properties. In lens applications the material, with reasonable care on the part of the user, performs very satisfactorily over periods of years.

Typical parts made of DAP.

DIALLYL PHTHALATE (DAP)

	PROPERTY	CR39-Cast sheet	Rogers 1-540	Rogers 520	Rogers 510N	Rogers 501AN	Plaskon 9-40-40	Plaskon 2-70-70VO	Durez 2761	Durez 4776
PHYSICAL	Mold shrinkage (in./in.)--in mils		1.5-3	1-3	3-6	7-10	2.5	2.5	6	5
	Specific gravity	1.32	2.08	1.75	1.66	1.78	1.72	1.91	1.71	1.78
	Specific volume (in.^{3}/lb)	21.0	13.3	15.83	16.69	15.56	16.1	14.5	16.2	15.56
	Water absorption (%) 24 hr @ 73F (23C)	0.2	0.25	0.25	0.5	0.4	0.35	0.34	0.2	0.2
	Water absorption (%) equilibrium, 73F (23C)	0.7								
	Haze (%)	0.3								
	Transmittance (%)	92								
MECHANICAL	Tensile strength (psi) yield									
	Tensile strength (psi) ultimate	$5\text{-}6\times10^{3}$	7,500	7,500	7,500	5,500			8,000	9,000
	Elongation (%) yield									
	Elongation (%) rupture	3								
	Tensile modulus (x10^5 psi)								14	13
	Flexural strength (psi)	7.5-8.5 M	13-16 M	15-18 M	12-15 M	10-13 M	13,500	10,250	14,000	13,000
	Flexural modulus (x10^5 psi)	2.5-3.5	24	17	13	11	13	13		
	Compressive strength (psi)	22,500	28,000	30,000	21,000	25,000			31,000	28,000
	Shear strength (psi)									
	Izod impact strength notched, 1/8 in.thick (ft-lb/in.)	0.2-0.4	0.7	1.0	0.45	0.46	4.3	1.1	0.5	0.5
	Izod impact strength unnotched, 1/8 in.thick (ft-lb/in.)	2-3								
	Tensile impact strength (ft-lb/in.^{2}) S type									
	Tensile impact strength (ft-lb/in.^{2}) L type									
	Fatigue strength (psi @ ___ cycles)									
	Rockwell hardness	M95-100								
	Deformation under load (%) @ 73F (23C) @ 66 psi									
	Deformation under load (%) @ 73F (23C) @ 264 psi									0.24
THERMAL	Heat deflection temperature (F)	140-190	400	505	325	300	500	>500	330	450
	Specific heat (Btu/lb/F or Btu/lb/C)	0.55								
	Thermal conductivity (Btu/hr/ft^2/F/in.)	1.45								
	Coeff. therm. exp.(x10^{-5})(in./in./F or /C)									
	Brittleness temperature (F)									
	Flammability UL Standard 94		94V-0	94HB		94HB		94V-0		94V-0
	Flammability Oxygen index							>30		
ELECTRICAL	Dielectric strength (V/mil) 50H	354	350	360	390	380	350	350		380
	Dielectric strength (V/mil) 10^{6}H									
	Dielectric constant 60H	4.4	4.3	2.7	3.7	3.7	4.2	4.6	4.6	4.6
	Dielectric constant 10^{6}H	3.6							4.4	4.4
	Power factor 60H	0.006	0.015	0.015	0.017	0.019	0.016	0.014	0.009	0.015
	Power factor 10^{6}H	0.041							0.014	
	Volume resistivity (ohm-cm)	4.1×10^{14}	10^{15}	10^{15}	10^{14}	10^{14}			10^{14}	10^{15}
	Arc resistance (sec), tungsten electrodes		180	150	135	135	130	130		

Diallyl Phthalate

Items 1 and 2 are glass-filled grades by Durez, a division of Hooker Chemicals, and Plastics Corp.

Items 3 and 4 are short glass filled and are products of Plaskon Products, Inc.

Item 5 is a mineral- and nylon-filled grade, *Item 6* is mineral-filled, and *Items 7 and 8* are short glass filled and are manufactured by Rogers Corporation.

Additional grades are produced by all suppliers to meet specific needs of military communications and similar applications.

The guiding factors for selection of these materials are electrical characteristics, impact strength (toughness), water absorption, heat resistance, compression strength, and processibility. The grades listed in the data sheet can be injection, transfer, and compression molded, except for *Item 5* which is suitable for compression and transfer molding. The choice of process will depend to a large degree on shape, dimensions, and other requirements, which may be inherent to any one of them.

Item 9 is cast sheet that broadly speaking belongs to the allyl family of plastics. It possesses outstanding mar and scratch resistance, along with good optical properties, and for those reasons it is used as a lens material for eye glasses, where weight and shattering are considerations. In comparison with glass, it weighs one-half as much. It is a product of PPG Industries.

The data and illustrations have been adapted from the commercial publications of the suppliers.

ALKYD

Alkyds are thermosetting resins. Chemically, according to the definition of ASTM-D883, the alkyd resin is a polyester convertible into a cross-linked form, requiring a reactant of functionality higher than two or having double bonds. This definition covers the "alkyd" as well as the "unsaturated polyester" also called the "thermoset polyester" resins. These resins are closely related, but commercially they are supplied by different companies and, as a rule, serve different types of customers. The major difference lies in the amount of cross-linking monomer present. The "alkyd" receiving a lesser amount in the compound determines the impact strength. Most of "alkyd" applications do not require Izod impact values above 0.5 ft-lb/in., whereas those of unsaturated polyester may call for impact strengths up to 40 times higher; therefore, they will require much more of the cross-linking monomer.

Alkyd—Granular or Nodular Form

The term "alkyd" used in above discussion refers to a raw material, prior to molding, in a granular or nodular form and, when molded, having normally an impact strength of 0.5 ft-lb/in. or less.

In "alkyds" the ingredients are essentially a polyester resin, a cross-linking agent, a filler, and inhibitors to retard cross-linking until the resin is to be used by the molder. Any one of the ingredients can be varied not only as to amount, but also as to kind of basic materials, as well as fillers, with the results that compounds can be formulated with a large range of properties. When one or more properties are optimized, it usually happens at the sacrifice of other properties.

Alkyd formulations display good dielectric strength at high temperatures and good arc and tracking resistance. Specially formulated grades are suitable as x-ray barriers, others may contain magnetizable ferrite. As a rule they have good flammability ratings. They have a tendency to pick up moisture in high humidity and steam conditions, which is a disadvantage in comparison with other thermosets. They have reasonably good colorability and good mechanical strength. The specific gravity ranges from 1.7 to 2.2, meaning relatively few cubic inches per pound.

Applications can be found in automotive ignition parts, appliance components, and electrical devices where arc resistance is a factor, such as switches and circuit breakers. The material can be processed by compression, transfer and injection molding.

Alkyd—Bulk or Extruded Form

The bulk or extruded alkyd refers to a raw material, prior to molding, in relatively large sizes and, when molded, having high impact strengths— up to 20 ft-lb/in., mostly due to fiberglass reinforcement. The sizes or slugs are usually large enough to fill the needs of a cavity when compression molded.

These materials are also called unsaturated polyester or thermosetting polyester.

In the thermoset polyester branch, the resin itself (without filler) is formulated from rigid to flexible and a multitude of grades between the two extremes. Considering additives, reinforcements and extenders (clay, chalk, calcium carbonates), and resin combinations, one can visualize a broad range of materials with properties to suit many applications.

Applications can be found in furniture, lamps, table tops, bowling balls, small boats (by layup or sprayup techniques), all-terrain vehicle bodies, custom auto bodies, truck cabins, playground equipment, trailers, etc. Reinforced sheet molding compounds (SMC) are compression molded for welding helmets, tote boxes, switch gear bases, automotive parts, bathtubs, and similar items. The bulk molding compound (BMC) is used for small parts in automobile and electrical components. In 80% of the applications fiberglass is used as a reinforcing filler.

Processing includes compression, transfer and injection molding, casting, hand layup, sprayup, vacuum bag molding, encapsulation, pultrusion, and centrifugal casting. Each process usually involves special formulations for the intended purpose. Broadly speaking, it is a versatile material, and the determining guidelines are cost of material and cost of processing.

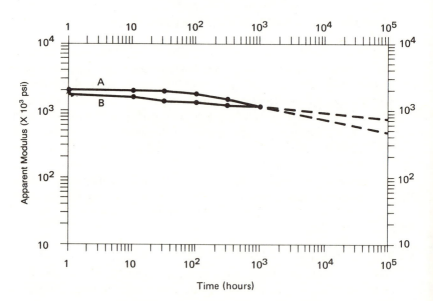

A, Glaskyd–1904 at 73°F (23°C) and 2000 psi; B, Glaskyd–1904 at 73°F (23°C) and 4000 psi. Dotted lines indicate extrapolated values.

ALKYD

PROPERTY		Cyglas 508	Cyglas 501	Glaskyd 01	Glaskyd 04	Glaskyd 02	Plenco 00	Durez 276	Durez 150	Durez 337
PHYSICAL										
Mold shrinkage (in./in.)--in mils		0	0–1	3–6	1–3	2–4	8–9	6	6–7	6–7
Specific gravity		1.82	1.89	2.16	2.14	2.05	1.86	1.9	2.2	1.98
Specific volume (in.^{3}/lb)		15.2	14.66	12.82	12.94	13.5	14.9	14.58	12.59	13.98
Water absorption (%) 24 hr @ 73F (23C)		0.06	0.06	0.22	0.1	0.2	0.31	0.2	0.1	0.3
Water absorption (%) equilibrium, 73F (23C)										
Haze (%)										
Transmittance (%)										
MECHANICAL										
Tensile strength (psi)	yield									
	ultimate	6–8 M	3–5 M	5–8 M			5,900	5,000	9,000	6,000
Elongation (%)	yield									
	rupture									
Tensile modulus (x10^5 psi)								13.0	28.0	19.0
Flexural strength (psi)		19–23 M	8–10 M	12–18 M	14–18 M	21–26 M	7,800	10,000	15,000	9,000
Flexural modulus (x10^5 psi)		24	20	30	30	20	12			
Compressive strength (psi)		18–22 M		20–24 M			19,200	22,000	35,000	25,000
Shear strength (psi)										
Izod impact strength notched, 1/8 in.thick (ft-lb/in.)		8–10	1.5–2	1.5–3	2–3	8–12	0.30	0.45	0.36	0.31
Izod impact strength unnotched, 1/8 in.thick										
Tensile impact strength (ft-lb/in.^{2}) S type										
L type										
Fatigue strength (psi @ ___ cycles)										
Rockwell hardness										
THERMAL										
Deformation under load (%) @ 73F (23C) @ ___ @ ___ F @ 66 psi										
@ 264 psi										
Heat deflection temperature (F)		>500	380	>500	>500	>440	360	400	450	375
Specific heat (Btu/lb/F or Btu/lb/C)										
Thermal conductivity (Btu/hr/ft^2/F/in.)										
Coeff. therm. exp.(x10^{-5}) (in./in./F or /C)		2–3	2–3	2	1.8					
Brittleness temperature (F)										
Flammability UL Standard 94		SE-0 1/8	SE-0 1/8					94V-0 @ 0.058	94V-0 @ 0.02	94HB @ 0.24
ELECTRICAL										
Dielectric strength (V/mil)		350	330	330	350	300	350	350	350	350
Dielectric constant 60H		4.6	5.2	6.7	6.3	7.1	6.2	5.1	5.2	6.7
Dielectric constant 10^{6}H		4.3	4.8	6.0	6.1	6.2	5.92	4.6	5.1	6.1
Dissipation factor 60H		0.05	0.02	0.01	0.01	0.03		0.01	0.011	0.02
Dissipation factor 10^{6}H		0.03	0.01	0.01	0.01	0.02		0.01	0.012	0.01
Volume resistivity (ohm-cm)				10^{15}				10^{14}	10^{16}	10^{14}
Arc resistance (sec), tungsten electrodes		>180	>180	>180	>180	150	180	185	184	180

Alkyd

Items 1–3 are granular materials by Durez, a division of Hooker Chemicals and Plastics Corporation. *Item 4* is also a granular, general purpose grade by Plenco, a tradename of Plastics Engineering Co.

Items 6 and 7 are "Glaskyd" and *Items 8 and 9* are "Cyglas" extruded or bulk grades of high impact alkyd materials. These are tradenames of American Cyanamid Co.

The Glaskyd compounds have a diallyl phthalate resin base, whereas in the Cyglas compounds, a polyester–styrene resin base is used. Each resin is used to achieve specific end use requirements.

There are additional grades available in the granular material; in the extruded and bulk materials, the number of grades and suppliers is quite large. The guiding factors for selecting a granular grade are electrical characteristics, impact strength, heat distortion, and rigidity.

Item 5 is mineral- and glass-reinforced grade formulated for maximum impact and flexural strengths.

Item 6 is also mineral and glass reinforced and was developed for applications where the requirement is good dimensional stability and creep resistance up to 300°F (149°C).

Item 7 is mineral-filled and glass-fiber-reinforced grade with excellent electrical properties and creep resistance.

Items 8 and 9 are mineral-filled and glass-reinforced grades for the electronic industry with high impact strength and excellent electrical properties in which surface and internal cracks as well as the tendency for internal voids are minimized.

The creep tests were conducted on a specimen in tension. Data and other information were adapted from the manufacturer's commercial publications.

Automotive ignition part made of granular alkyd.

Circuit-breaker base made of bulk alkyd.

AMINO

The aminos are thermosetting resins. Chemically, the resins can be described as being made by polycondensation of a compound containing amino groups, such as urea or melamine, with an aldehyde, such as formaldehyde, or an aldehyde-yielding material.

The most important groups are *melamine formaldehyde* and *urea formaldehyde.*

The major ingredients of these materials are not derived from petroleum. The products made from these molding compounds display the following characteristics: a high surface hardness with excellent mar resistance, a broad range of colorability, good heat resistance (a high heat deflection temperature), good flame resistance, and superior electrical arc and track resistance. They resist breakage and chipping, have good load-bearing strength, are rigid, and resist organic solvents, greases, and moisture. They are tasteless and odorless.

The melamines will outperform the ureas in heat- and chemical-resistance durability when exposed to boiling water and detergent solution; they are also available with a greater variety of fillers.

Some grades may show postmolding shrinkage extended over a prolonged period of time and, associated with this shrinkage, poor dimensional stability. This negative property can be overcome by appropriate fillers.

The highest percentage of melamines and ureas are filled with purified alpha-cellulose filler for improved strength, improved moldability, improved shrinkage characteristics, and good colorability. Other fillers are cotton flock, wood flour, fiberglass, macerated fabrics of nylon, cotton, and linen. Each one of the fillers or their combination causes an increase in such properties as impact strength, heat resistance, flexibility, etc. There also are grades of laminated moldings for industrial and decorative purposes.

The principal uses for urea molding compounds are wiring devices, bottle closures, housings for small controls and small appliances, and knobs and handles for electrical appliances.

A large portion of melamines goes into manufacture of dinnerware. Decorations are permanently molded on the desired surface by opening the mold prior to full cure and placing the treated decoration in the appropriate place, closing the mold, and, finally, curing the complete assembly. Other melamine applications include utensil handles, electric shaver housings, appliance parts, and similar products.

The processes for both materials are compression, transfer, and, with recently formulated grades, injection molding.

The resins are used as adhesives and paint finishes that provide a hard scratch-resisting coating.

A plug, receptacle, switch, and plate made
of urea formaldehyde.

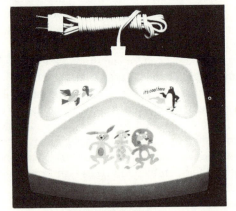

An electrically heated baby food dish made
of melamine.

A melamine–phenolic switch unit.

AMINO

PROPERTY	Beetle	Plaskon	Cymel 1077	M2015	M6204	M2880	Plenco Cellulose	Plenco -electr.
PHYSICAL								
Mold shrinkage (in./in.)—in mils	8-10		8-10	3-5	3-5	1-2	6	6
Specific gravity	1.5	1.5	1.5	1.5	1.72	1.93	1.64	1.65
Specific volume (in.^{3}/lb)	18.47	18.47	18.47	18.47	16.1	14.35	16.89	16.79
Water absorption (%) 24 hr @ 73F (23C)	0.4-0.8	0.6	0.4	0.5	0.2	0.4	0.43	0.35
equilibrium, 73F (23C)								
Haze (%)								
Transmittance (%)								
MECHANICAL								
Tensile strength (psi) yield / ultimate	5.5-7 M	7,000	7,500	7,000	5,000	8,000	6,500	6,500
Elongation (%) yield / rupture								
Tensile modulus ($\times 10^5$ psi)			13.5					
Flexural strength (psi)	11-18 M	12,000	13,500	14,000	8,000	18,000	8,500	9,300
Flexural modulus ($\times 10^5$ psi)	13-16	13	11	17	19	20	13	13
Compressive strength (psi)	30-38 M	30,000	42,500	32,000	26,000	30,000	26,500	23,800
Shear strength (psi)	11,500		11,500					
Izod impact strength notched,1/8 in.thick (ft-lb/in.) unnotched,1/8 in.thick	.27-.34	0.3	0.32	1.0	0.45	5.5	0.34	0.36
Tensile impact strength (ft-lb/in.^{2}) S type / L type								
Fatigue strength (psi @ __ cycles)								
Rockwell hardness	E94-97	M118	M120	M125	M115	M125		
Deformation under load (%) @ 73F (23C) @ __ F @ 66 psi @ 264 psi								
THERMAL								
Heat deflection temperature (F)	266	275	361	395	350	440	305	355
Specific heat (Btu/lb/F or Btu/lb/C)								
Thermal conductivity (Btu/hr/ft^2/F/in.)								
Coeff. therm. exp.($\times 10^{-5}$)(in./in./F or /C)								
Brittleness temperature (F)								
Flammability UL Standard 94 / Oxygen index		94V-0						
ELECTRICAL								
Dielectric strength (V/mil)	330-370	350	285	225	340	300	275	300
Dielectric constant 60H	7.8	7.9	8.9	8.0	5.6	6.2	9.32	15.06
Dielectric constant 10^6H	6.8	6.6	8.0				6.73	7.0
Dissipation factor 60H	0.036	0.04	0.07	0.07	0.03	0.02		
Dissipation factor 10^6H	0.03	0.03	0.03					
Volume resistivity (ohm-cm)	3×10^{11}		1.4×10^{12}					
Arc resistance (sec), tungsten electrodes	100-135	120	130	140	180+	180+	150	185

Amino

There are several grades in urea molding compounds whose properties in the molded condition are within the range listed in the data sheet. The difference is in the moldability characteristics as required by various types of products which apply them. Special grades are available for closures, wiring devices and small housings, switch plates (wall), buttons, and knobs. The complexity and size of the molded part determines what grade is selected.

Beetle is the trademark of American Cyanamid Co.; Plaskon is the trademark of Plaskon Products Inc.

Items 1 and 2 show the urea properties of the above companies.

Item 3 is the most widely used grade of melamine and is a trademark of American Cyanamid Co.

Items 4, 5, and 6 are grades of the Fiberite Corporation.

Item 4 is cellulose fabric filled, is an impact grade available in colors, and is applied to impact dishware, electrical components, housings, circuit breakers, junction boxes, and serving trays.

Item 5 is mineral-filled grade for conditions where arc resistance, heat resistance, flame resistance, and dimensional stability are required for such products as automotive and aircraft ignition parts, circuit breakers, switch gear components, and junction boxes.

Item 6 is a glass-fiber-reinforced grade that combines high impact strength with arc and track resistance, flame resistance, and dimensional stability for applications such as connectors, meter housings, and circuit breakers.

Items 7 and 8 are products of Plastic Engineering Co. and combine melamine and phenolic resin to provide the color stability of melamines and the moldability of phenolics. These compounds are suitable for switching devices and other products where good electrical insulation, abrasion resistance, and/or colorability are requirements.

The data and illustrations were adapted from commercial publications of suppliers.

CELLULOSIC

These plastics are based on cellulose compounds such as esters (cellulose acetate) and ethers (ethyl cellulose). Cellulose material is derived from wood pulp and cotton linters; it is a natural polymer. The thermoplastic cellulosic materials are produced by chemical modification of the natural cellulose. The commercial compounds vary in moldability and other properties depending on the percentage of plasticizers and other additives.

There are four basic grades available, with many subdivisions in each. All of them display certain common characteristics: They are hard, rigid, strong, tough, and transparent with unlimited color possibilities. As the molding flow of some compositions improves, the above properties decrease, while impact resistance increases. The dimensional stability is good; the water absorption is low; they resist aliphatic hydrocarbons and ethers. When exposed to low-molecular-weight alcohols, esters, ketones, aromatic hydrocarbons, and chlorinated hydrocarbons, they will swell or dissolve.

Electrically, they display good dielectric strength, good volume resistivity, and high dissipation factor. They can be formulated as "self-extinguishing," although they are generally rated as "slow burning."

Acetate

Acetate is used for such products as tool handles, extruded tape, and electrical appliance housings.

Butyrate

Butyrate is easier to process and tougher than acetate. It is used for pen and pencil barrels, signal lenses, and signs.

Propionate

Propionate is normally tougher and has higher strength and hardness than butyrate. It is used for automotive parts, blister packaging, and tooth brushes.

Ethyl Cellulose

Ethyl cellulose is not as resistant to acids as the preceding grades, but is more resistant to bases. This grade lends itself to formulations that meet requirements for higher heat resistance, high impact strength, and contact with food. Electrically, the dissipation factor and dielectric constant are low, but the dielectric strength is high. This grade is used for flashlight cases, electrical appliance parts, and fire extinguisher parts.

Cellulosic thermoplastic materials can be processed by injection molding and extrusion.

CELLULOSIC

PROPERTY		1. Acetate 036-MH	2. Acetate 043-MS	3. Butyrate 205-M	4. Butyrate 525-MH	5. Propionate 309-H4	6. Propionate 307-H
PHYSICAL	Mold shrinkage (in./in.)--in mils	5-8	5-8	3-6	3-6	3-6	3-6
	Specific gravity	1.27	1.28	1.19	1.2	1.21	1.2
	Specific volume (in.^{3}/lb)	21.8	21.64	23.28	23.08	22.89	23.08
	Water absorption (%) — 24 hr @ 73F (23C)	2.3	2.2	1.5	1.6	2.0	1.9
	equilibrium, 73F (23C)						
	Haze (%)						
	Transmittance (%)						
MECHANICAL	Tensile strength (psi) — yield	3,800	3,600	3,800	4,300	4,900	3,600
	ultimate	5,250	4,800	5,050	5,550	5,700	4,500
	Elongation (%) — yield						
	rupture						
	Tensile modulus (x10^5 psi)						
	Flexural strength (psi)	5,600	5,500	5,400	6,200	6,500	4,400
	Flexural modulus (x10^5 psi)	2.55	2.5	1.85	2.0	2.45	1.95
	Compressive strength (psi)	4,300	4,200	3,600	4,200	5,800	4,400
	Shear strength (psi)						
	Izod impact strength — notched,1/8 in.thick (ft-lb/in.)	4.0	3.7	5.3	4.2	3.0	8.3
	unnotched, 1/8 in.thick						
	Tensile impact strength (ft-lb/in.^{2}) — S type						
	L type						
	Fatigue strength (psi @ ___ cycles)						
	Rockwell hardness	R87	R82	R63	R79	R90	R51
	Deformation under load (%) @ ___ @ 73F (23C)						
	@2000 @ 122 F	13	20	14	7.0	1.0	21.0
THERMAL	Heat deflection temperature (F) @ 66 psi						
	@ 264 psi	134	131	143	155	175	136
	Specific heat (Btu/lb/F or Btu/lb/C)						
	Thermal conductivity (Btu/hr/ft^2/F/in.)						
	Coeff. therm. exp.(x10^{-5}) (in./in./F or /C)						
	Brittleness temperature (F)						
	Flammability — UL Standard 94						
	Oxygen index						
ELECTRICAL	Dielectric strength (V/mil)						
	Dielectric constant — 60H						
	10^{6}H						
	Power factor — 60H						
	10^{6}H						
	Volume resistivity (ohm-cm)						
	Arc resistance (sec), tungsten electrodes						

Screwdriver handle and ballpoint pen barrel.

Cellulosic

Items 1–6 are "Tenite" acetate, butyrate, and propionate materials. "Tenite" is the tradename of Eastman Plastics. Each material type is made in many grades and in addition to this, each grade is subdivided into an average of eight flows. The materials having varying flow numbers (indicated by designations after the dash) are provided to aid in the processing of parts with diverse configurations and cross sections. It should be emphasized that practically all properties change not only with grade number, but also with flow number within each grade.

Item 1 is suitable for cutlery handles, toilet seats, screwdriver handles and furniture hardware.

Item 2 is usually applied to combs, mechanical pencil parts, and closures.

Item 3 is normally applied for steering wheels, warning light lenses, and extruded sheet.

Item 4 is used for buttons, knobs, screwdriver handles, dial covers, name plates, sunglass frames, and tail light lenses.

Item 5 is applied for pen parts, tooth brush and hair brush handles, flashlight cases, camera parts, knobs, and goggles.

Item 6 is used for telephone parts, pen parts, ophthalmic frames, shoe heels, vacuum cleaner parts, face shields, and cosmetic items.

When a water absorption test is made with the specimen submerged for 24 hours at 73°F (23°C), there is a loss of soluble matter, with the amount varying with grade and flow number.

Any additional information on the above materials can be found in *Technical Reports* and *Material Bulletins* of Eastman Plastics.

Data and illustrations have been adapted form suppliers' literature.

Eyeglass frames.

EPOXY

Epoxies are thermosetting resins. Chemically, they are described as containing ether or hydroxyalkyl repeating units or both, resulting from the ring-opening reaction of lower-molecular-weight polyfunctional oxirane resins or compounds, with catalysts or various polyfunctional acidic or basic coreactants.

When cured, the epoxy resins show outstanding dimensional stability, high mechanical strength, excellent electrical properties, and good resistance to heat and chemicals.

The epoxies when filled with extenders and/or reinforcing materials—because of their outstanding adhesive properties to almost any material—enhance their characteristics to a degree not found in any other plastic material. Epoxy resins require a hardener to bring about a cured and rigid material.

Epoxy resin—glass laminates are used for electrical printed circuit boards, as structural members for airplanes, for pressure vessels, and tooling jigs—fixtures. Reinforcing materials in addition to glass fibers are cotton, paper, powdered metal, graphite, and synthetic fibers.

Casting, potting, and encapsulating compounds are used extensively for insulating electrical parts. When cured and solidified, they are free of voids, provide good electrical breakdown resistance, exhibit a low power factor, are resistant to moisture and chemicals, and have mechanical toughness. The castings, pottings, and encapsulations are made by pouring the resin-hardener composition into a mold which supports the electrical component that is to be insulated and which is cured in place. Fillers and extenders are used to reduce cost and minimize shrinkage, thus preventing cracks and voids in the finished product. This method of insulation is applied to electronic coils, switches, small transformers, as well as motors and generators.

Molding compounds consist of resin, hardener, and filler and additives formulated to provide good storage life prior to molding.

Molded parts are used for electronic components, as bobbins for coil windings, and in miscellaneous products where strength and other outstanding properties are a requirement.

The processes employed are compression, transfer, and injection molding.

Epoxy adhesives can be applied in connection with all types of material to be joined. They have a distinct advantage since they require a minimal amount of pressure for a satisfactory bond.

EPOXY

PROPERTY		1. Araldite MY720; hardener HT976 XV205	2. Resin XV 235; hardener XV205	3. Araldite 6010; hardener XV205	4. E-2748	5. E-9405	6. E-9451
PHYSICAL							
Mold shrinkage (in./in.)--in mils					4-6	5-7	4-6
Specific gravity					1.72	1.63	1.93
Specific volume (in.^{3}/lb)					16.1	16.99	14.35
Water absorption (%) 24 hr @ 73F (23C)					0.1	0.07	0.06
equilibrium, 73F (23C)							
Haze (%)							
Transmittance (%)							
MECHANICAL							
Tensile strength (psi) yield							
ultimate		8,500	8,000	12,000	11,000	10,000	9,000
Elongation (%) yield							
rupture		2	1.5	5.0			
Tensile modulus (x10^5 psi)		5.5	5.5	4.0			
Flexural strength (psi)		13,000	13,000	23,000	17,000	19,000	16,000
Flexural modulus (x10^5 psi)		5	4.5	5	14	15	20
Compressive strength (psi)		34,000			28,000	25,000	30,000
Shear strength (psi)							
Izod impact strength notched, 1/8 in.thick					0.7	0.4	0.3
(ft-lb/in.) unnotched, 1/8 in.thick							
Tensile impact strength (ft-lb/in.^{2}) S type							
L type							
Fatigue strength (psi @ ___ cycles)							
Rockwell hardness					M115	M115	M115
Deformation under load (%) @ ___ @ 73F (23C)							
@ ___ @ ___ F							
THERMAL							
Heat deflection temperature (F) @ 66 psi							
@ 264 psi		460	360	300	310	295	330
Specific heat (Btu/lb/F or Btu/lb/C)							
Thermal conductivity (Btu/hr/ft^2/F/in.)							
Coeff. therm. exp.(x10^{-5}) (in./in. /C)					1.8	2.4	2.3
Brittleness temperature (F)							
Flammability UL Standard 94							
Oxygen index							
ELECTRICAL							
Dielectric strength (V/mil)					400	400	
Dielectric constant 60H							
10^{6}H					4.5	3.6	3.8
Dissipation factor 60H							
10^{6}H					0.01	0.012	0.01
Volume resistivity (ohm-cm)					9x10^{15}	4x10^{15}	6x10^{15}
Arc resistance (sec), tungsten electrodes					180	140	180

Epoxy

Items 1–3 are two-component materials that are mixed and poured into a mold to encapsulate an electrical product. The finished product is a solid casting, which provides good electrical protection. These raw materials are products of Ciba-Geigy Corp.

Items 4–6 are molding compounds in which epoxy is the base material compounded with additives for use in processing finished parts. *Item 4* is applied to terminal strips, connectors, parts for textile processing equipment, missile components, and structural parts. It is a glass-reinforced, medium impact stength grade, suitable for molding complex shapes with good electrical, heat resistance, and chemical resistance properties.

Items 5 and 6 are used for encapsulation by the injection or transfer process of coils, resistors, capacitors, and other electronic components. In these items the electrical and mechanical needs dictate the selection of the grade. *Item 5* is a glass-fiber-filled material with excellent thermal shock, good moldability properties, and retention of electricals under humid conditions. *Item 6* is a mineral-filled material with a good balance of electrical and mechanical properties along with low water absorption and high thermal conductivity.

All the molding compounds can be postcured for a prescribed time and temperature and thereby appreciably increase their heat deflection temperature. The molding compounds are products of Fiberite Corporation. The data and descriptive information were adapted from the suppliers' commercial data sheets.

Numerous other grades from other suppliers are available.

FLUOROPLASTICS

Fluoroplastics are thermoplastic. These plastics are based on polymers made with monomers composed of fluorine and carbon only. They display outstanding characteristics in properties and, as such, it would be reasonable to expect that they be also very expensive ($4 to $25 per pound).

This family of materials has a useful temperature range −4°F to 500°F (−20°C to +260°C). When exposed to either cryogenic or elevated temperatures around 300°F (150°C), most of them will retain relatively good mechanical and electrical properties. Their electricial insulation properties are outstanding; their dielectric constant is low; the power factor is good, and the volume and surface resistivity are high; for all practical purposes they are chemically inert; they will not support combustion in air nor propagate flames when ignited; they pass the most extreme flammability tests; their nonadhesive properties and low coefficient of friction are unique. These materials are frequently used as an additive in thermosetting and thermoplastic compounds to improve the coefficient of friction. They can be filled with glass fibers, carbon graphite, or bronze to improve their physical properties while still retaining their inherent characteristics.

Whenever the enumerated properties are needed, the fluoroplastics can be used in such applications as liners for chemical processing in vessels, piping, and fittings; insulation for high-temperature wires, connectors, and wiring devices; flame-retarding laminates for aircraft interiors; nonstick coatings for cooking utensils; for piston rings, packings, "O" rings, and bearings in all types of equipment; and, finally, as electrical tape, which is also used for pipe threads to provide leak proof joints in extreme temperature conditions. The following grades are commercially available.

Electronic components.

Pump components.

FLUOROPLASTICS

PROPERTY	1. Teflon-TFE granular resin	2. Teflon-TFE fine powder	3. Teflon-FEP	4. Teflon-TFE (15% GF)	5. Teflon-TFE (15% graphite)	6. Teflon-TFE (60% bronze)	7. Tefzel 200, 280	8. Tefzel HT2004	9. FEP-LF1004	10. Tefzel-UF1006	11. Halar ECTFE 300, 500	12. Kel-F-PCTFE-81
PHYSICAL												
Mold shrinkage (in./in.)--in mils									2-4	4-6		
Specific gravity	2.2		2.14	2.22	2.19	3.97	1.70	1.86	2.21	1.89	1.68	0.96
Specific volume (in.^{3}/lb)	12.59		12.94	12.48	12.65	6.98	16.29	14.89	12.53	14.66	16.49	29
Water absorption (%) 24 hr @ 73F (23C)	<0.01	<0.01	<0.01	0.015		0.019	0.029	0.022	0.01	0.02	<0.1	0.05
Water absorption (%) equilibrium, 73F (23C)												
Haze (%)												
Transmittance (%)												
MECHANICAL												
Tensile strength (psi) yield							6,500	12,000		14,000	4,500	
Tensile strength (psi) ultimate	1.0-1.4 M	2.5-3.5M	3,000	2.8-3.6M	1.3-2.7M	1.8-2M			6,000		7,000	5,725
Elongation (%) yield								8			5	
Elongation (%) rupture	100-200	300-600	300	325	185	85	200	12	2-3	5	200	150
Tensile modulus (x10^5 psi)	0.5-0.9	0.4-0.9	0.95				1.2				2.4	20.0
Flexural strength (psi)	600			570	860	1,140			10,500	19,000	7,000	8,500
Flexural modulus (x10^5 psi)				3.12	2.03	1.97	2.0	9.5	8.0	10.5	2.4	1.8
Compressive strength (psi) @1%				1,000	1,080	1,120	7,100	10,000	5,500	11,500		5,500
Shear strength (psi)							6,000	6,500		7,000		
Izod impact strength notched,1/8 in.thick (ft-lb/in.)	2		2.9	2.7	2.6	2.0	No br.	9	8	7.5	No br.	5
Izod impact strength unnotched,1/8 in.thick (ft-lb/in.)									17	17-18		
Tensile impact strength (ft-lb/in.^{2}) S type												
Tensile impact strength L type												
Fatigue strength (psi @ ___ cycles)												
Durometer hardness	D50/65	D50/65	D55	D54	D55	D65	R50	R74			R93	D77
Deformation under load (%) @1000 @73F (23C)	15	2.4	1.8	11	9	5						
Deformation under load (%) @2000 @73 F												
THERMAL												
Heat deflection temperature (F) @ 66 psi	250						220	510	350	460	240	258
Heat deflection temperature (F) @ 264 psi							165	410			170	167
Specific heat (Btu/lb/F or Btu/lb/C)	0.25		0.28				0.46					0.215
Thermal conductivity (Btu/hr/ft^2/F/in.)	1.7			2.6	3.1	3.3	1.65	1.66				1.0
Coeff. therm. exp.(x10^{-5}) (in./in./F or /C)	5.5		5.2	3.8	4.7	4.5	5.2	0.9	2.4	1.6		
Brittleness temperature (F)												
Flammability UL Standard 94							94V-0	94V-0		94V-0	94V-0	94VE-0
Oxygen index												
ELECTRICAL												
Dielectric strength (V/mil)	600	600	600								490	
Dielectric constant 60H	2.1	2.1	2.1	2.4			2.6	3.4			2.6	
Dielectric constant 10^{6}H											2.5	
Dissipation factor 60H	0.0003		0.0007	0.009			0.005	0.005			<0.0009	
Dissipation factor 10^{6}H											0.009	
Volume resistivity (ohm-cm)	>10^{18}	>10^{18}	>10^{18}	>10^{13}	10^4		>10^{16}	>10^{16}			10^{15}	

FLUOROPLASTICS (continued) PROPERTY		13. Kynar PVDF (GRADE)
PHYSICAL	Mold shrinkage (in./in.)--in mils	3
	Specific gravity	1.77
	Specific volume (in.^{3}/lb)	15.65
	Water absorption (%) 24 hr @ 73F (23C)	
	equilibrium, 73F (23C)	
	Haze (%)	
	Transmittance (%)	
MECHANICAL	Tensile strength (psi) yield	6,300
	ultimate	
	Elongation (%) yield	
	rupture	
	Tensile modulus (x10^5 psi)	2.06
	Flexural strength (psi)	
	Flexural modulus (x10^5 psi)	8.7
	Compressive strength (psi)	2,100
	Shear strength (psi)	
	Izod impact strength notched, 1/8 in.thick	3-10
	(ft-lb/in.) unnotched, 1/8 in.thick	
	Tensile impact strength (ft-lb/in.^{2}) S type	
	L type	
	Fatigue strength (psi @ ___ cycles)	
	Durometer hardness	D80
	Deformation under load (%) @ ___ @ 73F (23C)	
	@ ___ @ ___ F	
THERMAL	Heat deflection temperature (F) @ 66 psi	234-284
	@ 264 psi	176-194
	Specific heat (Btu/lb/F or Btu/lb/C)	
	Thermal conductivity (Btu/hr/ft^2/F/in.)	0.8
	Coeff. therm. exp.(x10^{-5})(in./in./F or /C)	6
	Brittleness temperature (F)	
	Flammability UL Standard 94	94V-0
	Oxygen index	44
ELECTRICAL	Dielectric strength (V/mil)	250
	Dielectric constant 60H	8.4
	10^{6}H	6.1
	Dissipation factor 60H	0.049
	10^{6}H	0.016
	Volume resistivity (ohm-cm)	2x10^{14}
	Arc resistance (sec), tungsten electrodes	

Fluoroplastics

The headings in the text identify: the grade, tradename, and manufacturing company.

Items 1–8 are Du Pont's "Teflon" and "Tefzel" grades. Additional information and data on the Du Pont materials can be found in their *"Teflon"—Mechanical Design Data* and *"Tefzel"—Design Handbook*. The illustrations shown below were also adapted from the above Du Pont publications.

Item 9 is a 20%-glass-filled and chemically coupled FEP grade. *Item 10* is 30%-fiberglass-filled "Tefzel." These two items are manufactured by LNP Corp. Publications from this company are available on the subject of wear and friction characteristics of their compounds.

The chemical coupling of the fiberglass in *Item 9* significantly improves the mechanical properties while retaining the outstanding electrical, thermal aging, and chemical resistance properties of the base polymer.

Item 11 possesses outstanding barrier properties, especially to oxygen, carbon dioxide, chlorine gas, or hydrochloric acid in addition to other favorable characteristics found in fluoropolymers. It is a product of Allied Corp.

Item 12 is described as being "crystallizable," but never completely crystalline.

The "quick quenched" material is called "amorphous," and the "slow cooled" material is called "crystalline." These terms are relative, since only thin parts (0.070″) can be cooled rapidly to inhibit crystalline growth. The "amorphous" forms are less dense, more elastic, optically clear, and exceptionally tough. Thicker parts are less transparent, have higher tensile stength and lower elongation, greater resistance to the penetrations of liquids and vapors, and 392°F (200°C) use temperature. Additional information can be found in the *Engineering Manual on Kel-F81 Plastic* by the 3M Co.

There are additional grades on the market from the above and other suppliers.

Tensile strength vs temperature by ASTM D638; ASTM #5 Bars.

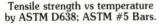

Tensile modulus vs temperature Tefzel® *HT-2004* ASTM D638; ASTM #5 Bars.

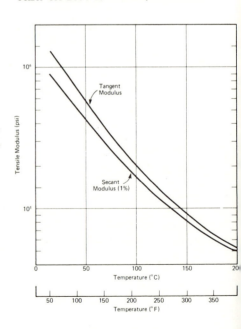

Creep—apparent flex modulus vs time and temperature by ASTM D674, Tefzel® *HT-2004.*

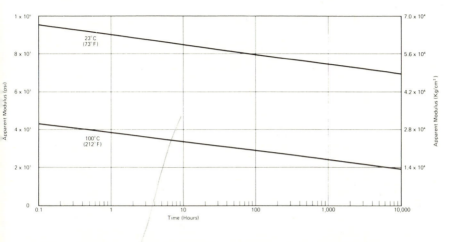

Tensile and compressive stress vs strain. By ASTM D638 at room temperature. Tensile specimens: ASTM #5. Compression specimens: cylinders 0.5″ diam. by 1″ long.

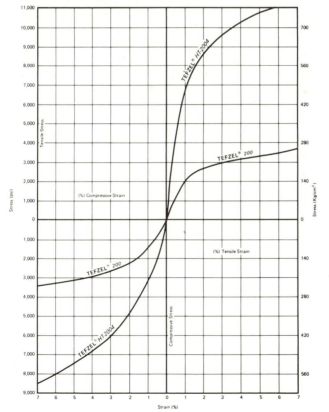

Stress vs strain in tension and compression (ASTM D695-54).

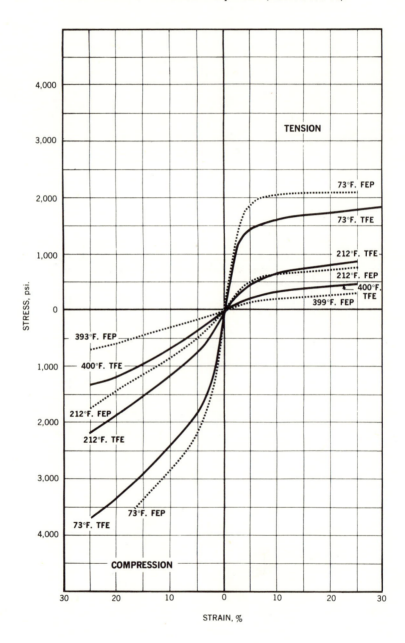

Polytetrafluoroethylene (PTFE) (Allied Chemical, "Halon"; DuPont Co., "Teflon"; I.C.I., "Fluon")

PTFE possesses properties on the high side of those described, along with high impact strength; however, with respect to tensile creep and wear resistance characteristics, they are low compared to other thermoplastics.

Filled PTFE is used for bearing pads in bridges and high rise buildings, for bushings, for seals in compressors, and for automotive transmission and hydraulic components.

The processing consists of compressing the powder, followed by sintering, similar to the method employed in powder-metallurgy. This procedure is necessary owing to the very high melt temperatures needed that are not attainable in conventional equipment.

Fluorinated ethylene propylene copolymer (FEP) (DuPont Co., "Teflon")

The service temperature of FEP is about 390°F (200°C). The tensile strength, creep resistance, and wear properties are also relatively low. In most other categories the properties and applications are nearly comparable to PTFE.

FEP can be injection molded, extruded, transfer molded, or blow molded. The only precaution necessary is that the processing facilities be protected against corrosion by the use of such materials as Xaloy or Duranickel when the plastic is at or near melt temperature.

Perfluoralkoxy resin (PFE) (DuPont Co., "Teflon")

PFE retains useful mechanical properties over the range of −328 to 500°F (−200 to +260°C). In other respects the properties are similar to PTFE and FEP. The processing includes injection, transfer, and blow molding and extrusion. Here again precautions are necessary against corrosion, in same manner as for FEP.

Polychlorotrifluoroethylene (PCTFE) (3M Co., "Kel-F")

PCTFE's chemical resistance is somewhat inferior to the previously described grades. This material will be dissolved by some solvents at 212°F (100°C), and some other solvents can cause swelling. Most other properties are similar to the Teflon grades. PCTFE can be optically clear in thicknesses up to ⅛″. It also possesses outstanding resistance to permeation of water vapor and gases in comparison with other plastics. This characteristic is used to advantage in the film applied to blister packaging of medicinal tablets and capsules. Other applications are generally the same as those described in previous grades. PCTFE can be injection molded and extruded on conventional thermoplastic equipment, thermoformed, heat-sealed, and vacuum metallized and printed. Its film can also be laminated.

Ethylene—chlorotrifluoroethylene copolymer (ECTFE) (Allied Chemical Corp., "Halar")

The temperature resistance of ECTFE is in the neighborhood of 356°F (180°C), and may be extended up to 392°F (200°C) for short duration. Its chemical resistance is good against organic solvents as well as corrosive chemicals at room and also at elevated temperatures. The mechanical properties are superior to those of FEP. Other characteristics fall into the range of the fluoroplastics in general.

Applications are wire and cable coatings used in mass transit cars, power plants, lighting fixtures, and appliance wire. Molded shapes cover packings, seals, "O" rings, and valve parts. Tubing is used as liners for hose for laboratories and as insulation.

The material is available for processing by injection molding, extrusion, rotomolding, fluidized bed coating, and electrostatic coating. The processing equipment should have anticorrosive material, such as Xaloy and Duranickel, wherever the material is at or near melt temperature.

Ethylene—tetrafluoroethylene copolymer (ETFE) (DuPont Co., "Tefzel")

The chemical resistance, electrical properties, and flammability of ETFE approach those of the Teflon materials, but its mechanical strength is superior. The useful temperature is on the order of 356°F (180°C).

Processing can be by injection molding or extrusion, and the equipment has to resist the corrosive action of the material similarly to the other compounds. Applications also fall in the same general category as other Teflons.

Polyvinylidene fluoride (PVDF) (Pennwalt Co., "Kynar")

The useful properties of PVDF extend up to 302°F (150°C). The mechanical strength, creep resistance, and wear compare favorably with the fluoroplastic copolymers. Applications are similar to the other materials described except that this grade is additionally used as a base for high-quality finishes for exterior siding.

Processing can be by injection molding, extrusion, transfer molding, and for coating either in dry powder form or dispersion.

Polyvinyl fluoride (PVF) (Dupont Co., "Teflon")

PVF is supplied as a tough, yet flexible, film. The film has useful properties up to 302°F (150°C); it has good weathering properties and resists abrasion. PVF can be laminated to a variety of plastics, metals, wood, and other materials and can serve as the exposed covering on the inside of aircaft and in the building field.

FURAN

Furan is a thermosetting plastic. In furan resins the furan ring is an integral part of the polymer chain, the furan being the largest amount by mass. Furan plastics are not based on petroleum, but are derived from renewable sources such as vegetable byproducts (corn cobs, oat hulls, etc.). Economics should stimulate extensive research into this material in order to extend its usefulness. It has good chemical resistance against most acids, alkalis, and solvents. The resin will harden in a few seconds at temperatures around 500°F (260°C). When reinforced with glass fibers, it can be used as ducts and liners for processing equipment as a protection against the corrosive action of various chemicals. At present the material is not available in a form that would be of general interest to a product designer.

IONOMER

Ionomer is a thermoplastic. Chemically, it can be described as having ethylene as its major component, but containing both covalent and ionic bonds. The anions hang from the hydrocarbon chain, and the cations are metallic—magnesium, zinc. The long molecular chains of the polymer have ionic cross-links that give the material a partial degree of the stability found in cross-linked materials. There are many grades made in the ionomer family of varying properties to suit specific uses. This variation is accomplished by changing the proportions of carboxylic acid with respect to the amount and type of metallic ion.

Ionomer has good resilience, flexibility, elongation, abrasion resistance, resistance to cutting, and maintaining a constant modulus of elasticity in the range of temperatures from −40°F to 100°F (−40°C to 40°C). Good flexibility exists even at very low temperatures. Additives in the polymer are needed for UV stability; other weatherability resistance, such as moisture permeability and water absorption, is good. Clarity and colorability are features that can be advantageous to its utilization. Chemical resistance at room temperature is also good.

The applications of ionomer include covers of golf balls to provide cut resistance, in the shoe industry for various components for foot protection, and in the automotive field for bumper pads. Film grade is coextruded with other materials as packaging material for food and drugs, as well as for blistering purposes. It provides ease of sealing along with resistance to puncture and tearing. Ionomer can be processed by injection molding and extrusion.

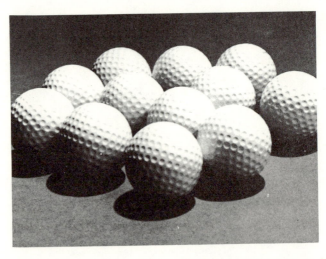

The toughness and impact resistance of Surlyn ® is demonstrated by its outstanding performance as a golf ball covering.

IONOMER

PROPERTY	GRADE	1. Surlyn 1554	2. Surlyn 1559	3. Surlyn 1605	4. Surlyn 1650	5. Surlyn 1702	6. Surlyn 1855
PHYSICAL							
Mold shrinkage (in./in.)--in mils							
Specific gravity		0.95	0.94	0.95	0.95	0.95	0.96
Specific volume (in.^{3}/lb)		29.16	29.47	29.16	29.16	29.16	28.85
Water absorption (%) 24 hr @ 73F (23C)							
equilibrium, 73F (23C)							
Haze (%)		26	6	5	15	7	7
Transmittance (%)							
Tensile strength (psi) yield		1,800	1,800	2,300	1,600	1,600	
ultimate		3,700	4,200	4,800	4,100	3,600	3,800
Elongation (%) yield							
rupture		410	450	470	460	460	510
Tensile modulus (x10^5 psi)							
Flexural strength (psi)							
Flexural modulus (x10^5 psi)		0.38	0.32	0.51	0.30	0.28	0.14
Compressive strength (psi)							
Shear strength (psi)							
Izod impact strength notched,1/8 in.thick		10.1	11.4	19.2	No br.	No br.	No br.
(ft-lb/in.) unnotched, 1/8 in.thick							
Tensile impact strength (ft-lb/in.^{2}) S type							
L type		570	550	485	590	360	610
Fatigue strength (psi @ ___ cycles)							
Rockwell hardness							
Deformation under load (%) @ ___ @ 73F (23C)							
@ ___ @ ___ F							
Heat deflection temperature (F) @ 66 psi		108	111	111	104	104	104
@ 264 psi							
Specific heat (Btu/lb/F or Btu/lb/C)		0.43	0.43	0.43	0.43	0.43	0.43
Thermal conductivity (Btu/hr/ft^2/F/in.)		5.8	6.0	5.9	6.0	5.8	5.7
Coeff. therm. exp.(x10^{-5}) (in./in. /C)		15	14	10	16	14	17
Brittleness temperature (F)							
Flammability UL Standard 94							
Oxygen index							
Dielectric strength (V/mil)							
Dielectric constant 60H							
10^{6}H							
Power factor 60H							
10^{6}H							
Volume resistivity (ohm-cm)							
Arc resistance (sec), tungsten electrodes							

Ionomer

Surlyn is a tradename of the Du Pont Co.

The number of grades is twice that shown on the sheet; they possess a combination of several outstanding features as required for specific applications.

Item 1 is most abrasion resistant.

Item 2 has best low-temperature toughness.

Item 3 is the stiffest.

Item 4 has the best melt adhesion.

Item 5 is the clearest.

Item 6 has the most flex crack resistance.

Data and illustration adapted from Du Pont's publication on Surlyn.

NITRILE RESIN

Nitrile resins are thermoplastic copolymers and consist of the following monomers (by weight): 70% acrylonitrile, 20−30% methylacrylate or styrene, and 0−10% butadiene. It is a lightweight, impact resistant, transparent material with outstanding resistance to permeation of vapors and gases and good resistance to chemicals, such as dilute acids and bases, esters and ketones, liquid and chlorinated hydrocarbons. The mechanical properties are also favorable. With these properties it is an excellent material for the packaging of all types of liquids, being very similar to glass. Nitrile resin can also be applied to parts in instruments or instruments or meters that are exposed to water vapors or gases where low permeability could prevent potential problems such as condensation. It can be injection molded, blow molded, extruded, or thermoformed.

	NITRILE RESIN		GRADE	1. Packaging grade
	PROPERTY			
PHYSICAL	Mold shrinkage (in./in.)--in mils			2-5
	Specific gravity			1.15
	Specific volume (in.³/lb)			24.09
	Water absorption (%)	24 hr @ 73F (23C)		0.28
		equilibrium, 73F (23C)		
	Haze (%)			
	Transmittance (%)			
MECHANICAL	Tensile strength (psi)	yield		
		ultimate		9,000
	Elongation (%)	yield		
		rupture		3-4
	Tensile modulus (x10⁵ psi)			5.1
	Flexural strength (psi)			
	Flexural modulus (x10⁵ psi)			5.1
	Compressive strength (psi)			12,000
	Shear strength (psi)			
	Izod impact strength notched, 1/8 in.thick			4.0
	(ft-lb/in.) unnotched, 1/8 in.thick			
	Tensile impact strength (ft-lb/in.²)	S type		
		L type		
	Fatigue strength (psi @ ___ cycles)			
	Rockwell hardness			M72
	Deformation under load (%) @ ___ @ 73F (23C)			
	@ ___ @ ___ F			
THERMAL	Heat deflection temperature (F) @ 66 psi			172
	@ 264 psi			164
	Specific heat (Btu/lb/F or Btu/lb/C)			
	Thermal conductivity (Btu/hr/ft²/F/in.)			
	Coeff. therm. exp.(x10⁻⁵)(in./in. /C)			6
	Brittleness temperature (F)			
	Flammability	UL Standard 94		
		Oxygen index		
ELECTRICAL	Dielectric strength (V/mil)			220
	Dielectric constant	60H		
		10⁶H		
	Power factor	60H		
		10⁶H		
	Volume resistivity (ohm-cm)			
	Arc resistance (sec), tungsten electrodes			

NYLON

Nylon is a thermoplastic resin. Nylon is the generic name for a group of resins that consist of long chain polymeric amide, which has recurring amide groups as an integral part of the main polymer chains. They are also called polyamide plastics.

Mechanical properties can vary over a broad range. All nylons are hygroscopic and in molded or extruded condition will absorb moisture until equilibrium is reached. At this point the material increases in toughness, dimensions increase, and its modulus is lowered. *During design of a product, the effects of moisture on properties should be an important consideration.*

As a group the nylons display high strength, toughness, good fatigue resistance, resistance to most industrial chemicals, a low coefficient of friction, and abrasion resistance. The creep resistance and resistance to deformation at elevated temperature are good. The rigid grades show a high degree of notch sensitivity. For applications requiring exposure to sunlight for long periods of time, nylons containing UV stabilizers are used, the most effective being carbon black. Nylons are generally inert and are not attacked by fungi or bacteria. Since its resistance to permeability of oxygen is good, nylon is suitable for many packaging applications. The electrical characteristics are also good. When requirements for a product are established, the likelihood of finding an appropriate grade of nylon is favorable.

The applications are extensive in the automotive field for wire connectors, windshield wiper parts, speedometer parts, timing sprockets, radiator fans, steering column components, etc. For electrical purposes they are used for coils, bobbins, wiring devices, and instrument parts, and all sorts of small appliances. Miscellaneous uses are for sporting goods; for spatulas, spoons, and serving knives in the kitchen; and for bristles in paint and tooth brushes. Industrially, nylon is used for cams, gears, bearings, rollers on machinery. Nylons can be processed by injection, extrusion, and most other thermoplastic processing means.

There are several groups in the nylon family and each group may consist of many grades for specific uses. One of the basic groups, prepared from a diamine and a diacid, is 66, 69, 610, and 612. The 66 is the most widely used material, whereas the remainder of the group are specialty resins and are considerably more expensive.

The other basic group, prepared from a lactam, is 6, 11, and 12. Type 6 polycaprolactam is the second most widely used nylon. Type 11 has very low moisture absorption and can be plasticized to be suitable for automotive and industrial hose applications. Type 12 is the lowest moisture-absorbing grade of any commercial nylon available. Types 11 and 12 are considerably more expensive than the widely used 6 and 66 grades.

In addition to the enumerated grades, copolymerization between grades is possible with resulting properties far different than those of the components.

ADE

NYLON

PROPERTY	Zytel-101 (6)-DAM	Zytel-101 (6)-50% RH	Zytel 08 HS (66) 60% RH	Zytel 158L (12)-DAM	Zytel ?-801-DAM	Minlon 0B-DAM	Zytel GRZ-G13L-DAM	Zytel GRZ-G33L-DAM	Zytel GRZ-G43L-DAM
PHYSICAL									
Mold shrinkage (in./in.)--in mils	15			11	15	3	11	6	2
Specific gravity	1.14	1.14		1.06	1.08	1.42	1.18	1.35	1.46
Specific volume (in.³/lb)	24.3	24.3		26.13	25.65	19.51	23.47	20.52	18.97
Water absorption (%) 24 hr @ 73F (23C)	1.2			0.25	1.2			0.5	0.14
Water absorption (%) equilibrium, 73F (23C)	8.5		1.09	3.0	6.7		6.1	4.6	1.7
Haze (%)									
Transmittance (%)									
MECHANICAL									
Tensile strength (psi) yield	12,000	8,500	7,500	8,800	7,500	12,200	14,000	22,000	28,000
Tensile strength (psi) ultimate	12,000	11,200	7,500	8,000					
Elongation (%) yield	5	25	15	7		3	4	3	4
Elongation (%) rupture	60	≥300	270	150	60				
Tensile modulus (x10⁵ psi)	4.1	5.0	1.6	2.95	2.45	7.0			
Flexural strength (psi)							21,000	33,000	39,000
Flexural modulus (x10⁵ psi)							5.5	10.0	15.0
Compressive strength (psi)	4,900			2,400			19,000	29,000	23,000
Shear strength (psi)	9,000			8,600	8,400		4,000	10,500	12,000
Izod impact strength notched, 1/8 in. thick	1.0	2.1	1.2	1.0	17.0	1.1	2.3	2.4	2.5
Izod impact strength unnotched, 1/8 in. thick									
Tensile impact strength (ft-lb/in.²) S type	75	110	120	73	280				
Tensile impact strength L type	240	700	800	291					
Fatigue strength (psi @ __ cycles)									
Rockwell hardness	M79	M59	M50	R114	R112	R120	M82	M96	R118
THERMAL									
Deformation under load (%) @ 73F (23C) @ 66 psi / @2000 @122 F @ 264 psi	1.4			1.0		@4000...	1.7	1.3	0.5
Heat deflection temperature (F) @ 66 psi	455			356	421	496			
Heat deflection temperature (F) @ 264 psi	194			194	160	473	450	475	410
Specific heat (Btu/lbF or Btu/lb/C)	0.4			0.4					
Thermal conductivity (Btu/hr/ft²/F/in.)	1.7			1.5					
Coeff. therm. exp.(x10⁻⁵) (in./in./F or /C)	4			5		2	1.3	1	1.2
Brittleness temperature (F)	-112	-85	-120	-190					
Flammability UL Standard 94	94V-2	94V-2	94HB	94V-2	94HB				
Flammability Oxygen index	28	31	20	25	18				
ELECTRICAL									
Dielectric strength (V/mil) 60H									
Dielectric strength 10⁶H									
Dielectric constant 60H	4.0	8	5.9	4.0	3.2			3.4	3.6
Dielectric constant 10⁶H	3.8	4.6	3.3	3.9	2.9				
Dissipation factor 60H	0.01	0.2	0.1	0.02	0.01			0.02	0.02
Dissipation factor 10⁶H	0.02	0.1	0.01	0.02	0.02				
Volume resistivity (ohm-cm)	10¹⁵	10¹³	10¹³	10¹³	10¹⁴			10¹⁴	10¹⁵
Arc resistance (sec), tungsten electrodes									

NYLON (continued)

PROPERTY	10. Vydyne 22H-(66)	11. Vydyne 80X-(66/6)	12. Vydyne 60H(69)	13. Vydyne R100-R200 (reinf.)	14. Vydyne R400G(GR)	15. Celanese Nylon 1003 (66)	16. Celanese Nylon 1310-1	17. Celanese Nylon 1310-4	18. Celanese Nylon 1500 (GR)
PHYSICAL									
Mold shrinkage (in./in.)--in mils	15-20	15-20	11-13	10-22		15-20			
Specific gravity	1.14	1.14	1.08	1.47	1.42	1.14	1.14	1.14	1.38
Specific volume (in.³/lb)	24.3	24.3	25.65	18.84	19.51	24.3	24.3	24.3	20.1
Water absorption (%) 24 hr @ 73F (23C)	1.1	1.6	0.44	0.6		1.5	1.4	1.4	1.0
equilibrium, 73F (23C)									
Haze (%)									
Transmittance (%)									
MECHANICAL									
Tensile strength (psi) yield	12,400	11,900	10,200	14,000	16,500	12,000	13,600	13,600	28,000
ultimate						12,000	13,600	13,600	28,400
Elongation (%) yield	10	11	10			40-80	35-55	35-55	
rupture	70	110	50	9	7				4
Tensile modulus (x10⁵ psi)	4.3	4.0	2.75	9.85	9.7	4.2			
Flexural strength (psi)	13,000	15,000	11,000	22,000	26,700	17,000	19,000	17,000	41,000
Flexural modulus (x10⁵ psi)	4.0	3.9	3.35	8.98	9.8	4.2	5.0	4.7	13.0
Compressive strength (psi)									29,400
Shear strength (psi)						9,500	9,400	9,400	13,000
Izod impact strength notched,1/8 in.thick (ft-lb/in.²)	1.0	1.0	0.7	1.0	1.0	1.0	0.85	1.0	2.2
unnotched,1/8 in.thick (ft-lb/in.)					10.5				
Tensile impact strength (ft-lb/in.²) S type	80	70	50	110					70
L type									
Fatigue strength (psi @ ___ cycles)									
Rockwell hardness	R120	R119	M61	R119	R120	M79	M86	M84	
Deformation under load (%) @ 73F (23C) @ ___ F									
THERMAL									
Heat deflection temperature (F) @ 66 psi	450	410	333	475	426	470	465	425	495
@ 264 psi	180	160	167	380		170	174	171	485
Specific heat (Btu/lb/F or Btu/lb/C)	0.4	0.4	0.4	0.35	0.37	0.4	0.4	0.4	0.3
Thermal conductivity (Btu/hr/ft²/F/in.)	1.7	1.7	1.5	3.12	2.7				1.5
Coeff. therm. exp.(x10⁻⁵)(in./in./F or /C)	4.5	4.5	8.3	2.7	1.7	5			1.6
Brittleness temperature (F)									
Flammability UL Standard 94	94V-2	94V-2	94HB	94V-2					
Oxygen index									
ELECTRICAL									
Dielectric strength (V/mil)	570	650	547	490		600	600	600	400
Dielectric constant 60H	3.1	3.6	3.2	3.9		5.3	5.2	5.2	4.0
10⁶H									
Dissipation factor 60H	0.03	0.03	0.02	0.02		0.035	0.27	0.027	
10⁶H									
Volume resistivity (ohm-cm)	10^{15}	10^{14}	10^{15}	10^{15}					10^{14}

NYLON (continued)

	PROPERTY	VF-1008	QFL 4536 Migralube	Q-1000 (unfilled)	RC-1008 & 40% carbon fiber filled	PFL-4216 30% FG	Fosta on 512(6)	Fosta on 523(6)	Fosta on 1417(6)
PHYSICAL	Mold shrinkage (in./in.)--in mils	5	3.5	13	2	4-5	12	12	26
	Specific gravity	1.38	1.45	1.08	1.34	1.43	1.14	1.14	1.1
	Specific volume (in.³/lb)	20.07	19.1	25.65	20.67	19.4	24.3	24.3	25.2
	Water absorption (%) 24 hr @ 73F (23C)	0.8	0.15	0.4	0.4	1.0	1.7	1.8	1.3
	equilibrium, 73F (23C)	4.5				6.0			
	Haze (%)								
	Transmittance (%)								
MECHANICAL	Tensile strength (psi) yield	20,000	20,000	8,500	40,000	18,500	12,500	12,500	9,000
	ultimate						9,900	7,900	6,300
	Elongation (%) yield	3-4	3-4	12	3-4	3-4	11	11	10.2
	rupture						20		150
	Tensile modulus (x10⁵ psi)								
	Flexural strength (psi)	27,000	29,000	12,000	60,000	28,000	17,000	16,000	10,000
	Flexural modulus (x10⁵ psi)	11.0	11.5	2.8	34.0	11.0	4.0	3.95	2.75
	Compressive strength (psi)	15,000			14,000	21,000	4,450	3,550	3,900
	Shear strength (psi)	6,700							
	Izod impact strength notched, 1/8 in. thick (ft-lb/in.)	3.5	2.2	0.6	1.6	1.6-2.0	0.8	0.8	3.0
	unnotched, 1/8 in. thick (ft-lb/in.)	20.0	20.0		13.0	12.0			
	Tensile impact strength (ft-lb/in.²) S type						170	170	260
	L type								
	Fatigue strength (psi @ cycles)								
	Rockwell hardness					M94	R120	R120	R113
	Deformation under load (%) @ 73F (23C) @ 2000 psi / @4000@122F					0.5			
THERMAL	Heat deflection temperature (F) @ 66 psi		410		500	425	375	375	367
	@ 264 psi			135		420	155	146	133
	Specific heat (Btu/lb/F or Btu/lb/C)								
	Thermal conductivity (Btu/hr/ft²/F/in.)	3.1		5	8.5				
	Coeff. therm. exp.(x10⁻⁵)(in./F or /C)	1.6	2.7		0.8	1.9			
	Brittleness temperature (F)								
	Flammability UL Standard 94	94HB				94HB	94V-2		94HB
	Oxygen index								
ELECTRICAL	Dielectric strength (V/mil)						495		468
	Dielectric constant 60H / 10⁶H								
	Dissipation factor 60H / 10⁶H								
	Volume resistivity (ohm-cm)								
	Arc resistance (sec), tungsten electrodes								

Adding fillers, reinforcements, extenders, and other additives to all these materials, we can readily visualize nylon grades that cover a broad range of properties that will meet innumerable applications.

Nylon

Items 1–9 are grades trademarked "Zytel" of the Du Pont Co.

Items 1 and 2 are a general-purpose 66 grade. Its properties are shown as DAM (dry as molded) and also when exposed to 50% relative humidity. The remaining items will have their properties shown mostly as DAM. At 50% relative humidity, the properties are more realistic. A direct comparison of grades can be realized easier when the DAM data are evaluated. For this reason, most of the data will be shown in the DAM condition. The direction of change in properties on all grades, as far as DAM or 50% relative humidity is concerned, will be the same, except that the size of change may vary from grade to grade.

Item 3 is a modified 66 resin for superior toughness, added impact resistance and flexibility, in addition to improved heat stability.

Item 4 is a 612 nylon. It has a much lower water absorption than the previous items without sacrificing other properties to any appreciable degree.

Item 5 is designated as a super tough grade with a very high impact resistance. In all other respects, it compares with properties of 66 nylons.

Item 6 is mineral or mineral/glass-reinforced nylon. This type of material displays uniform mold shrinkage, low part warpage, improved stiffness, and heat resistance.

Item 7 is 13% glass-reinforced lubricated grade. It is used where shock loading is encountered at elevated service temperatures and maximum toughness is required.

Item 8 is a 33% glass-reinforced lubricated grade. It has outstanding resistance to impact with only moderate reduction in mechanical and load-carrying properties that prevail in other grades.

Items 7 and 8 are based on impact modified 66 resin.

Item 9 is a low moisture absorption grade based on 612 resin. It is a 43% glass-reinforced lubricated grade. It has the highest dimensional stability of glass-reinforced nylons and also has excellent electrical properties over a wide range of humidities.

Items 10–14 are grades trademarked "Vydyne" of Monsanto Plastics and Resins Co. All properties shown are DAM and at 50% relative humidity change as shown in the illustrated curves.

Item 10 is a general-purpose, heat-stabilized, type-66 grade.

Item 11 is a copolymer of 66/6 nylons providing greater ductility at low moisture than 66 alone.

Item 12 is a 69 type, which provides a lower moisture absorption and good dimensional stability in comparison with type 66.

Item 13 is mineral reinforced for applications requiring high rigidity, high tensile strength, and high heat distortion temperature.

Items 15–18 are products of the Celanese Plastics and Specialties Co.

Item 15 is a general-purpose grade, heat stabilized for use in parts located under the hoods of automobiles.

Item 16 is similar to Item 15 and is suitable for parts that can benefit from higher rates of production due to their shape and cross sections.

Item 17 is of the same type as Item 16, except that it incorporates an additive for easy mold release.

Item 18 is a 33% glass-reinforced nylon that is formulated to have excellent impact strength, high tensile strength, and high modulus, along with a high use temperature of up to 350°F (177°C).

Items 19–21 are grades trademarked "Fosta" nylon of American Hoechst Corporation. They are type-6 nylons.

Item 19 is a general-purpose grade suitable for motor housings, seat belt buckles, and automotive fasteners.

Item 20 is suitable for bearings, bushings, valves, door slides, gears, and hardware items.

Item 21 is stiffer than Item 20 and can be used for pipe fittings, buttons, electrical connectors, and kitchen utensils.

Items 22–26 are grades trademarked "Thermocomp" glass fortified nylons by LNP Corp.

Item 22 is a type-6 nylon filled with 30% fiberglass and molybdenum disulfide lubricant to provide a balance of strength, toughness, abrasion, and wear resistance. The applications are for gears, cams, slides, bearings, and wherever unlubricated contact under load exists.

Item 23 is type-66 nylon fortified with 40% carbon fiber. The properties with this reinforcement are truly outstanding. In comparison with glass-reinforced nylon, this material has reduced mold shrinkage and thermal coefficient of expansion along with greatly increased thermal conductivity. It is a remarkable material.

Item 24 shows the properties of an unfilled 610 nylon that is being used as the base material for Items 25 and 26.

Item 25 called a "Migralube" resin. This item is filled with 30% fiberglass, 15% TFE, and silicone. The silicone is the "migratory" lubricant that forms a continuous self-renewing boundary at the surface of the part. The silicone in combination with the TFE forms an extreme pressure lubricant for severe frictional applications, exceeding the performance of previous formulations for such requirements.

Item 26 is the "super tough" impact-modified nylon filled with 40% fiberglass. The listed properties show potential applications for which plastics were not considered before.

The listed nylon items represent only selected grades from a large number of available types from the above mentioned as well as other sources.

The illustrations that follow, as well as data and information, have been adapted from the suppliers technical publications. For additional information on the behavior of nylon materials see: Du Pont, *"Zytel"—Design Handbook;* Monsanto, *"Vydyne" Resins—Part Design Manual;* Celanese Plastics, *Bulletins NIA & NIB;* American Hoechst Corp., *Bulletin—"Fosta" Nylon;* LNP Corp., *Bulletin 203–278.*

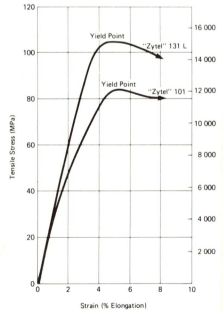

Tensile stress–strain data for
"Zytel" 101 and "Zytel" 131
dry as molded, at 73°F (23°C).

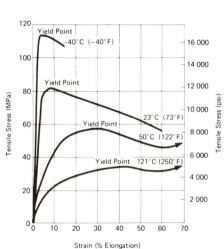

Tensile stress–strain data for
"Zytel" 101, dry-as-molded,
at four temperatures.

Tensile stress–strain data for "Zytel" 101, 50% RH at four temperatures.

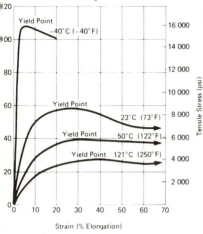

Stress–strain in tension and compression of "Zytel" 101, 73°F (23°C).

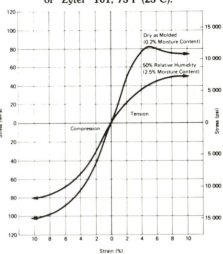

ote: Consider heat stabilized "Zytel" for long term exposure to high temperatures — e.g. "Zytel" 103 HS-L.

*Sample has been conditioned to 50% relative humidity, henceforth referred to as 50% RH.

Stress–strain curves in tension and compression at 73°F (23°C) (ASTM D638 and D695).

Vydyne R-100 & R-200

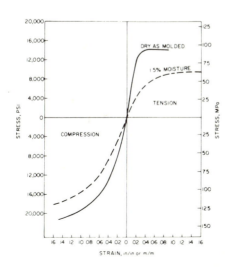

Tensile strength at break vs temperature (ASTM D638).
Vydyne R-100 & R-200

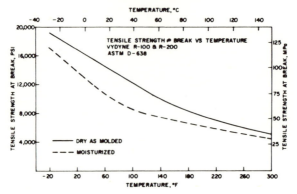

Dimensional change as a function of cyclic variations in environmental relative humidity.

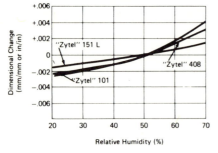

Moisture content vs time for "Zytel" 101 exposed to 50% RH air at 73°F (23°C).

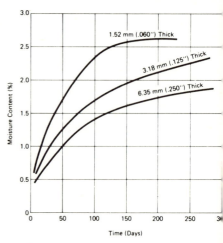

Typical postmolding changes due to stress-relief/moisture absorption, unannealed samples.

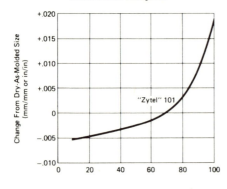

76.2 mm x 127.0 mm x 1.6 mm   % RH
(3" x 5" x 1/16") Plaques

Gate = 1/2 Part Width
       1/2 Part Thickness

Dimensional change with temperature*
unreinforced nylons.

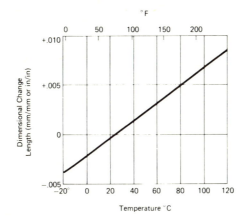

Coefficient of Linear Thermal Expansion assumed to be
5x10⁻⁵ in./in./°F over entire range.

Creep in flexure of "Zytel" 101, 73°F (23°C), 50% RH.

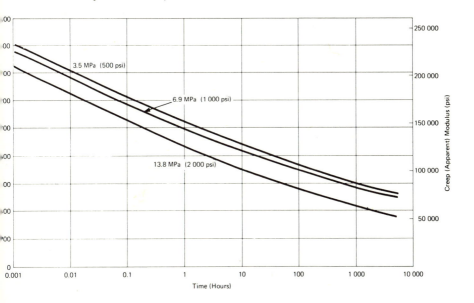

Tensile stress vs strain (ASTM D638).

**Vydyne R-100 & R-200
Dry-as-Molded**

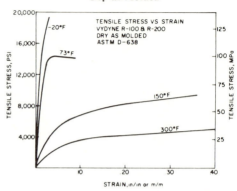

**Vydyne R-100 & R-200
at 1.5% Moisture Content**

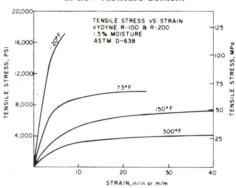

Coefficient of linear thermal expansion
vs temperature (ASTM D696).

Vydyne R-100 & R-200

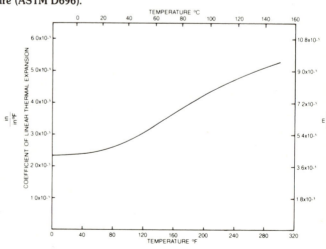

Tensile modulus vs temperature
(ASTM D638).
Vydyne R-100 & R-200

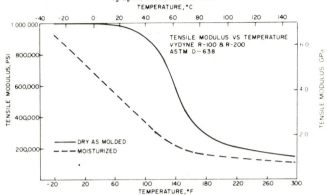

Tensile modulus vs moisture content
(ASTM D638-72).

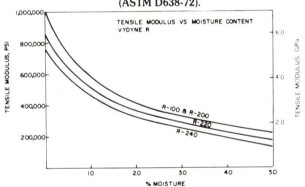

Tensile creep modulus vs time
[ASTM D674; 73°F (23°C) at
50% relative humidity].
Vydyne R-100 & R-200

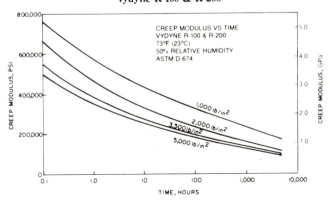

Apparent modulus vs
time at 73°F (23°C).

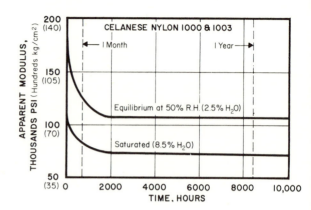

Apparent modulus vs
time.

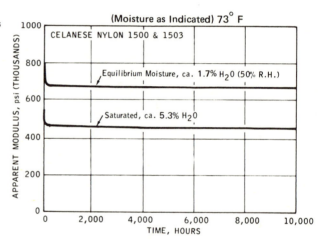

Dimensional change vs
moisture content.

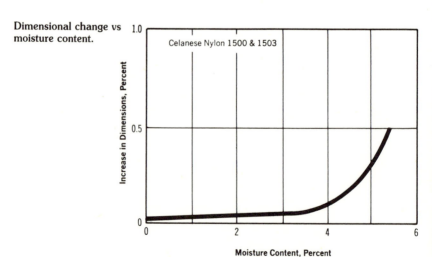

Percent dimensional change vs temperature.

Dimensional change of "as-molded" part in air exposure at 73°F (23°C) and 50% RH.

Dimensional change vs percent moisture.

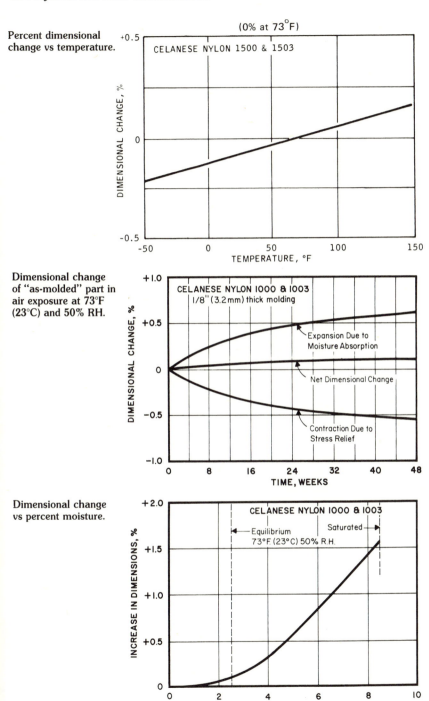

(0% at 73°F)

CELANESE NYLON 1500 & 1503

CELANESE NYLON 1000 & 1003
1/8" (3.2 mm) thick molding

Expansion Due to Moisture Absorption

Net Dimensional Change

Contraction Due to Stress Relief

CELANESE NYLON 1000 & 1003

Equilibrium
73°F (23°C) 50% R.H.

Saturated

Effect of temperature
on notched Izod
impact strength
(dry as molded).

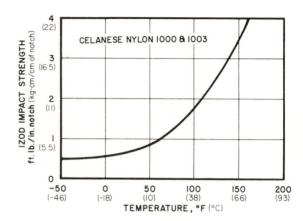

Effect of moisture content
on notched Izod
impact strength at
73°F (23°C).

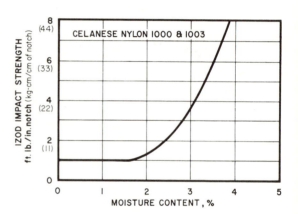

TENSILE ELONGATION AT BREAK VS. TEMPERATURE (ASTM D-638)
Vydyne R-100 & R-200

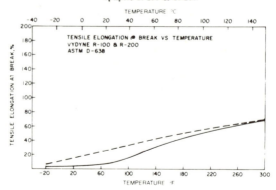

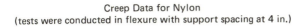

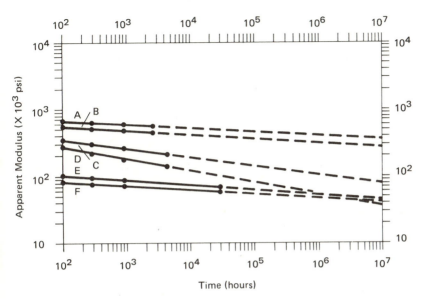

A, Zytel 70G–32 (33% glass reinforced) at 73°F (23°C) and 4000 psi; B, Zytel 70G–32 (33% glass reinforced) at 140°F (60°C) and 4000 psi; C, Vydyne R100 and R200 (mineral reinforced) at 73°F (23°C) and 1000 psi; D, Vydyne R100 and R200 (mineral reinforced) at 73°F (23°C) and 2000 psi; E, Nylon 6 at 73°F (23°C) and 1000 psi; F, Zytel 109 at 73°F (23°C) and 1000 psi. All materials were at equilibrium with 50% relative humidity. Dotted lines indicate extrapolated values.

These electrical connectors are sufficiently tough for use in industrial areas.

PHENOLIC

Phenolic resins and compounds are thermosetting. They are based on resins made by condensation of phenols, such as phenol cresol, with formaldehyde and other aldehydes. Their desirable properties made them the "workhorse" plastic for engineering and appliance applications over a long period of time (70 years or more).

The molded phenolic products display good heat resistance, favorable electrical properties, resistance to chemicals and moisture absorption, dimensional stability, relatively low creep characteristics, good rigidity, excellent reproducibility in size and shape, and flammability resistance. The compounds consist by volume of 35−50% resin and 50−65% fillers. The fillers can be wood flour, paper, cotton flock, chopped fabric, glass fibers, mica, etc., either individually or in combination with each other to impart certain properties such as high heat resistance, impact strength, minimal moisture absorption, nonbleeding, etc. The variation in resin properties along with various combinations of fillers provides a variety of grades that will match almost any design requirement.

The resins are used for molding compounds for such products as wiring devices, electrical switch gears, electrical relay systems, connectors, closures, etc. In the automotive industry they are used for brake systems, transmission parts, valves, etc. In household products they are used as handles and knobs for utensils and small appliances. The resin itself is used for laminated products in which fabrics, paper, or glass fibers are molded into boards that are applied to some product designs.

At present there are compounds suitable for processing by compression, transfer, and injection molding. The main drawback of the material was that the "excess" from production and from defective parts were a total loss. At present injection processing methods have been developed whereby "excess" material is eliminated, thus making the overall benefit of phenolic products more attractive.

A typical phenolic part.

PHENOLIC

PROPERTY	Fiberite M-9294	Fiberite M-7700	Fiberite 001	Plenco 509 electr.	Plenco 0-GP	Genal 4301E heat resistant	Genal 02-Impact	Genal 050-GP	Durez 5-Electr.	Durez 56-Impact	Durez 3-GP
PHYSICAL											
Mold shrinkage (in./in.)--in mils	1-3	2-4	2-4	3-7	5-10	7.5	9	12	4-5	6-9	7-9
Specific gravity	1.38	1.38	1.40	1.60	1.39	1.51	1.38	1.39	1.68	1.38	1.4
Specific volume (in.^{3}/lb)	20.07	20.07	19.79	17.31	19.93	18.34	20.07	19.93	16.49	20.07	19.79
Water absorption (%) 24 hr @ 73F (23C)	0.8	0.6	0.6	0.08	0.27	0.2	1.1	0.3	0.15	0.8	0.5
equilibrium, 73F (23C)											
Haze (%)											
Transmittance (%)											
MECHANICAL											
Tensile strength (psi) yield											
ultimate	9,000	6,100	5,800	9,700	8,500	7,000	7,000	6,500	5,500	6,500	8,000
Elongation (%) yield											
rupture											
Tensile modulus (x10^5 psi)						10.0	10.0	10.0	15.0	14.0	14.0
Flexural strength (psi)	16,000	12,500	8,500	15,000	14,000	11,000	10,000	9,000	9,000	10,000	11,000
Flexural modulus (x10^5 psi)	14.0	13.0	11.0	13.0	9.5						
Compressive strength (psi)	22,000	24,000	22,800	26,000	29,500	30,000	27,000	27,000	26,000	20,000	30,000
Shear strength (psi)											
Izod impact strength notched,1/8 in.thick (ft-lb/in.)	4.5	1.9	0.6	0.28	0.29	0.28	0.50	0.30	0.36	1.1	0.34
unnotched,1/8 in.thick (ft-lb/in.)											
Tensile impact strength (ft-lb/in.^{2}) S type											
L type											
Fatigue strength (psi @ cycles)											
Rockwell hardness	M110	M105	M113			M100	M114	M110			
THERMAL											
Deformation under load (%) @ 73F (23C) @ 56 psi											
@ 264 psi F											
Heat deflection temperature (F)	300	330	365	350	330	375	325	335	375	350	340
Specific heat (Btu/lb/F or Btu/lb/C)											
Thermal conductivity (Btu/hr/ft^2/F/in.)	9.0	9.0	9.0			6.0	4.8	5.6			
Coeff. therm. exp.(x10^{-5}) (in./in./F or /C)	1.3	2	2.4			2.5	3.2	2.7			
Brittleness temperature (F)											
Flammability UL Standard 94						94V-0	94HB	94HB			94V-0
Oxygen index											
ELECTRICAL											
Dielectric strength (V/mil)	320	300	300	300	270	300	325	340	375	350	400
Dielectric constant 60H	4.5	4.2	4.4	7.1	4.96	22.5	12.4	7.8	6.0	6.0	5.90
10^6 H									5.0	4.5	4.7
Dissipation factor 60H	0.05		0.05			0.25	0.5	0.2	0.08	0.16	0.10
10^6 H									0.06	0.04	0.05
Volume resistivity (ohm-cm)	2x10^{13}	2x10^{13}							10^{12}	10^{13}	10^{13}
Arc resistance (sec), tungsten electrodes						180			184		184

PHENOLIC (continued) PROPERTY	12. Fiberite FM-4030-190	13. Rogers RX842	14. Rogers RX468	15. Rogers RX630	16. Rogers RX862
PHYSICAL					
Mold shrinkage (in./in.)--in mils	0.6-1.2	1.2	2	1	1.5
Specific gravity	1.74	1.83	1.72	1.75	1.88
Specific volume (in.^{3}/lb)	15.92	15.14	16.10	15.83	14.73
Water absorption (%) 24 hr @ 73F (23C)	0.15	0.05	0.35	0.07	0.07
equilibrium, 73F (23C)					
Haze (%)					
Transmittance (%)					
MECHANICAL					
Tensile strength (psi) yield					
ultimate	7,500	6,000	6,000	12,000	6,500
Elongation (%) yield					
rupture					
Tensile modulus (x10^5 psi)					
Flexural strength (psi)	16,000	12,000	10,500	23,000	12,000
Flexural modulus (x10^5 psi)	24.0	17.0	18.0	22.0	23.0
Compressive strength (psi)	27,000	24,000	26,000	40,000+	28,000
Shear strength (psi)					
Izod impact strength notched, 1/8 in.thick	12.0	0.70	0.75	1.2	0.90
(ft-lb/in.) unnotched, 1/8 in.thick					
Tensile impact strength (ft-lb/in.^{2}) S type					
L type					
Fatigue strength (psi @ ___ cycles)					
Rockwell hardness	M115				
Deformation under load (%) @ ___ @ 73F (23C)					
@ ___ @ ___ F					
THERMAL					
Heat deflection temperature (F) @ 66 psi					
@ 264 psi	500+	430	500+	450	500
Specific heat (Btu/lb/F or Btu/lb/C)					
Thermal conductivity (Btu/hr/ft^2/F/in.)	10.0				
Coeff. therm. exp.(x10^{-5})(in./in./F or /C)	1.5				
Brittleness temperature (F)					
Flammability UL Standard 94		94V-0	94V-0	94V-0	94V-0
Oxygen index					
ELECTRICAL					
Dielectric strength (V/mil)	380				
Dielectric constant 60H					
10^{6}H	5.0				
Dissipation factor 60H					
10^{6}H	0.02				
Volume resistivity (ohm-cm)	10^{13}				
Arc resistance (sec), tungsten electrodes			175	180	183

Phenolic

Items 1–3 are "Durez" products, a trademark of Hooker Chemicals and Plastics Corporation.

Items 4–6 are "Genal" products, a trademark of General Electric Co.

Items 7–8 are "Plenco" products, a trademark of Plastics Engineering Co.

Items 9–12 are products of the Fiberite Corp. and *Items 13–16* are products of Rogers Corp.

Each supplier has a large number of additional compounds to meet the variety of end use needs. Additional information can be obtained from the manufacturers.

The guidelines for selecting a phenolic material are impact resistance, temperature of application, moisture absorption, and electric characteristics.

The data and illustrations have been adopted from technical bulletins on the materials.

PHENYLENE-OXIDE-BASED RESIN (NORYL)

Noryl is a thermoplastic resin. Polyphenylene oxide is produced by a patented process via oxidative coupling of a phenolic monomer. It is then blended with rubber-modified polystyrene to give an alloy which is called Noryl.

Molding products display excellent dimensional stability, low creep, high modulus, good mechanical properties over a broad range of temperatures, low water absorption, good hydrolitic stability, good electrical properties, and excellent impact strength. It is one of the best materials for faithful, repeated reproduction of dimensions. It is available in the following grades: (1) Flame-retardant grades with a high heat deflection temperature and a good stability under load at elevated temperatures; (2) a general purpose grade for miscellaneous applications; (3) platable grades; (4) glass-reinforced grades. And, as the need arises, new grades are being developed to satisfy market requirements.

Applications are for electrical appliances, business machines, automotive grilles, wheel covers, electrical connectors, plumbing parts, TV parts, etc.

Processing is done by injection and extrusion. It is one of the easiest materials to process, especially where close tolerance dimensions are involved.

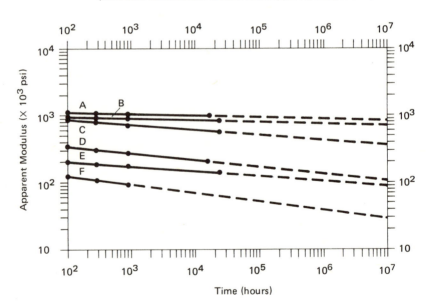

Creep Data for Noryl
(tests were conducted in flexure with support spacing at 4˙in.)

A, Noryl GFN3 at 73°F (23°C) and 2000 psi; B, Noryl GFN3 at 150°F (66°C) and 2000 psi; C, Noryl GFN3 at 170°F (77°C) at 2000 psi; D, Noryl 731 at 73°F (23°C) and 2000 psi; E, Noryl 731 at 150°F (66°C) and 2000 psi; F, Noryl 731 at 170°F (77°C) and 2000 psi. Dotted lines indicate extrapolated values

PHENYLENE-OXIDE-BASED RESIN

	PROPERTY		1. Noryl 731	2. Noryl SE-1	3. Noryl SE-100	4. Noryl GFN2	5. Noryl GFN3	6. Noryl PN235
PHYSICAL	Mold shrinkage (in./in.)--in mils		5-7	5-7	5-7	2-4	1-3	5-7
	Specific gravity		1.06	1.06	1.10	1.21	1.27	1.05
	Specific volume (in.^{3}/lb)		26.2	26.2	25.2	22.9	21.9	26.38
	Water absorption (%) 24 hr @ 73F (23C)		0.066	0.066	0.07	0.06	0.06	0.07
	equilibrium, 73F (23C)							
	Haze (%)							
	Transmittance (%)							
MECHANICAL	Tensile strength (psi)	yield						
		ultimate	9,600	9,600	7,800	14,500	17,000	7,000
	Elongation (%)	yield						
		rupture	60	60	50	4-6	4-6	
	Tensile modulus (x10^5 psi)		3.55	3.55	3.8	9.25	12.0	
	Flexural strength (psi)		13,500	13,500	12,800	18,500	20,000	9,800
	Flexural modulus (x10^5 psi)		3.6	3.6	3.6	7.5	11.0	3.0
	Compressive strength (psi)		16,400	16,400	12,000	17,600	17,900	
	Shear strength (psi)		10,500	10,500	6,900	10,400	10,400	
	Izod impact strength notched, 1/8 in.thick		5.0	5.0	5.0	2.3	2.3	5.0
	(ft-lb/in.) unnotched, 1/8 in.thick							
	Tensile impact strength (ft-lb/in.^{2})	S type						
		L type	170	170	150	60	50	
	Fatigue strength (psi @2x10^{6}ycles)		2,500	2,500	1,850	4,000	5,000	
	Rockwell hardness		R119	R119	R115	L106	L108	R114
	Deformation under load (%) @ @ 73F (23C) @2000@122 F		0.3	0.3	0.5	0.2	0.12	
THERMAL	Heat deflection temperature (F) @ 66 psi		279	279	230	293	317	250
	@ 264 psi		265	265	212	290	300	235
	Specific heat (Btu/lb/F or Btu/lb/C)							
	Thermal conductivity (Btu/hr/ft^2/F/in.)		1.5	1.5	1.10	1.15	1.10	
	Coeff. therm. exp.(x10^{-5}) (in./in./F or /C)		3.3	3.3	3.8	2	1.4	3.5
	Brittleness temperature (F)							
	Flammability UL Standard 94		SB	SE-1	SE-1	SB	SB	HB
	Oxygen index							
ELECTRICAL	Dielectric strength (V/mil)		550	500	400	420	550	450
	Dielectric constant 60H		2.65	2.69	2.65	2.86	2.93	2.61
	10^{6}H		2.64	2.68	2.64	2.85	2.92	
	Dissipation factor 60H		0.0004	0.0007	0.0007	0.0008	0.0009	0.0004
	10^{6}H		0.0009	0.0024	0.0024	0.0014	0.0015	
	Volume resistivity (ohm-cm)		10^{17}	10^{17}	10^{17}	10^{17}	10^{17}	
	Arc resistance (sec), tungsten electrodes		75	75	75	70	120	

Phenylene-Oxide-Based Resin (Noryl)

Noryl is a tradename of General Electric Co.

Item 1 is a general-purpose grade.

Item 2 is similar to *Item 1*, except that it has UL SE-1 rating and a UL 221°F (105°C) continuous use temperature approval. It is a suitable material for critical current carrying and sole support applications.

Item 3 also has UL SE-1 rating, but lower heat-deflection temperature. It is used for products where a lower heat deflection, UL rating, and excellent dimensional stability are the key requirements.

Items 4 and 5 are 20% and 30% glass-reinforced resins that provide high strength, rigidity, higher heat-deflection temperature, and toughness. Other properties such as creep resistance and lower coefficient of expansion are also more favorable and, yet, the precision of dimensional moldability is maintained.

Item 6 is a plating grade.

The data and illustrations have been adapted from *"Noryl" Design* of General Electric Co. where additional data on other grades and general information on varying conditions of application can be found.

Typical parts made with Noryl.

Noryl 731 stress–strain diagram—tensile.

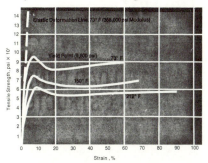

Apparent modulus, GFN3.

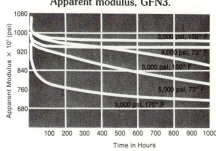

Tensile strength vs temperature.

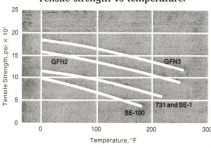

Flexural strength vs temperature.

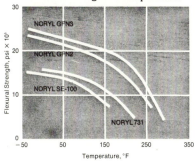

GFN-3 flexural creep behavior at elevated temperatures.

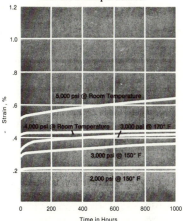

Apparent modulus, Noryl 731.

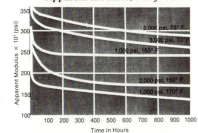

Tensile modulus vs temperature.

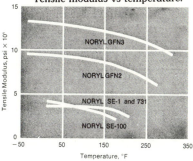

Flexural modulus vs temperature.

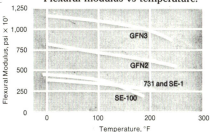

POLY(AMIDE–IMIDE) (TORLON)

Poly(amide–imide) is a thermoplastic material. It is a very high priced material with properties to match its cost. The mechanical properties are excellent, and precision molded parts will maintain their integrity when exposed continuously to temperatures up to 500°F (260°C). The modulus of elasticity is 660,000 psi and, when reinforced with graphite fiber, will reach 2.6×10^6 psi. The data sheet should be closely evaluated for other outstanding properties. The chemical resistance is good against aliphatic and aromatic hydrocarbons, halogenated solvents, and most acid and base solutions. High-temperature caustic, steam, and some acids will attack this plastic.

There are available a general purpose grade, a grade reinforced with 30% graphite fiber for high modulus, and a grade filled with PTFE/graphite for a low coefficient of friction, high resistance to creep and wear, high compressive strength, and a low coefficient of thermal expansion.

The applications are for critical components in business machines, automotive–aircraft–space industries, also for bearings, gears, pumps, and valves where its material characteristics are required.

The processing can be done by injection molding or extrusion.

These bearings and seals are examples of typical parts made with Torlon.

Poly(amide–imide)

"Torlon" is the tradename of Amoco Chemicals Corp.

Item 2 is a formulation for bearings.

Any additional information that is being developed can be obtained from the supplier of the material.

The data and illustration are adapted from the supplier's publications.

POLY (AMIDE-IMIDE)

		GRADE	1. Torlon 4203	2. Torlon 4301
	PROPERTY			
PHYSICAL	Mold shrinkage (in./in.)--in mils			
	Specific gravity		1.4	1.45
	Specific volume (in.^{3}/lb)		19.79	19.1
	Water absorption (%) 24 hr @ 73F (23C)		0.28	0.22
	equilibrium, 73F (23C)			
	Haze (%)			
	Transmittance (%)			
MECHANICAL	Tensile strength (psi) yield			
	ultimate		26,900	19,500
	Elongation (%) yield			
	rupture			
	Tensile modulus (x10^5 psi)			
	Flexural strength (psi)		30,700	26,900
	Flexural modulus (x10^5 psi)		6.64	9.6
	Compressive strength (psi)		32,000	30,000
	Shear strength (psi)		18,300	16,100
	Izod impact strength notched, 1/8 in.thick		2.5	1.1
	(ft-lb/in.) unnotched, 1/8 in.thick			
	Tensile impact strength (ft-lb/in.^{2}) S type			
	L type			
	Fatigue strength (psi @ ___ cycles)			
	Rockwell hardness			
	Deformation under load (%) @ ___ @ 73F (23C)			
	@ ___ @ ___ F			
THERMAL	Heat deflection temperature (F) @ 66 psi			
	@ 264 psi		525	525
	Specific heat (Btu/lb/F or Btu/lb/C)			
	Thermal conductivity (Btu/hr/ft^2/F/in.)		1.4	5.1
	Coeff. therm. exp.(x10^{-5}) (in./in./F or /C)		1.8	1.6
	Brittleness temperature (F)			
	Flammability UL Standard 94		94VE-0	94VE-0
	Oxygen index			
ELECTRICAL	Dielectric strength (V/mil)			
	Dielectric constant 60H			
	10^{6}H			
	Power factor 60H			
	10^{6}H			
	Volume resistivity (ohm-cm)			
	Arc resistance (sec), tungsten electrodes			

POLYARYL ETHER (ARYLON T)

Polyaryl ether is a thermoplastic material; it is an alloy of polysulfone and of a specialty grade ABS.

Polyaryl ether is best known for its high impact strength, which is *independent of its cross-sectional thickness*. It has high heat resistance combined with good chemical resistance. It can be readily plated as well as painted (even oven baked) with standard finishing materials. Hydrolysis resistance is another favorable property to the degree that it can be exposed to boiling water for prolonged times. The creep resistance is also good.

Chemically, it will resist organic solvents except chlorinated aromatics, esters, and ketones. It also displays good resistance to aliphatics and inorganic reagents.

The applications are in the automotive, appliance, and plumbing fields for use as mechanical and electrical components. It can be easily processed by injection molding, extrusion, and thermoforming. It can also be solvent welded.

Arylon T is a trade name of Uniroyal Chemicals.

POLYARYL ETHER

	PROPERTY		1. Arylon T- Uniroyal
PHYSICAL	Mold shrinkage (in./in.)--in mils		7
	Specific gravity		1.14
	Specific volume (in.^{3}/lb)		24.3
	Water absorption (%)	24 hr @ 73F (23C)	
		equilibrium, 73F (23C)	
	Haze (%)		
	Transmittance (%)		
MECHANICAL	Tensile strength (psi)	yield	
		ultimate	7,500
	Elongation (%)	yield	
		rupture	
	Tensile modulus (x10^5 psi)		3.2
	Flexural strength (psi)		12,000
	Flexural modulus (x10^5 psi)		3.3
	Compressive strength (psi)		
	Shear strength (psi)		
	Izod impact strength notched,1/8 in.thick		8
	(ft-lb/in.) unnotched, 1/8 in.thick		
	Tensile impact strength (ft-lb/in.^{2})	S type	
		L type	
	Fatigue strength (psi @ ___ cycles)		
	Rockwell hardness		R117
	Deformation under load (%) @ ___ @ 73F (23C)		
	@ ___ @ ___ F		
THERMAL	Heat deflection temperature (F)	@ 66 psi	
		@ 264 psi	300
	Specific heat (Btu/lb/F or Btu/lb/C)		
	Thermal conductivity (Btu/hr/ft^2/F/in.)		1.7-2.1
	Coeff. therm. exp.(x10^{-5}) (in./in./F or /C)		3.6
	Brittleness temperature (F)		
	Flammability	UL Standard 94	94HB
		Oxygen index	
ELECTRICAL	Dielectric strength (V/mil)		760
	Dielectric constant	60H	3-5
		10^{6}H	2.8-3.8
	Power factor	60H	0.004
		10^{6}H	0.006
	Volume resistivity (ohm-cm)		10^{15}
	Arc resistance (sec), tungsten electrodes		

POLYBUTYLENE

Polybutylene is a thermoplastic material; it is prepared by the polymerization of butene as the sole monomer. Polybutylene is one of the latest additions to the polyolefin family.

Polybutylene possesses high toughness, abrasion resistance, good flexibility, resistance to elevated temperatures [230°F (110°C)], good barrier properties, and resistance to mechanical and environmental stress cracking and to creep under load. It is an outstanding electrical insulating material, and is capable of accepting fillers to as high as 85% by weight.

Chemically, it will resist soaps, detergents, and most acids and bases below a temperature at 200°F (93°C) and most aliphatic solvents at room temperature. It is partially soluble in aromatic and chlorinated solvents above 140°F (60°C) and will absorb hydrocarbons at room temperature.

The applications are piping for plumbing (hot and cold water), sheeting and film for domestic and industrial applications, and as insulation for electric cables.

Processing is adaptable to extrusion, blow molding, injection molding, and rotomolding.

POLYCARBONATE (PC)

PC is thermoplastic material of outstanding characteristics. It is amorphous and is a polyester polymer in which the repeating structural unit in the chain is of the carbonate type.

The material has excellent properties in these categories: Impact strength, creep resistance, broad range of use temperature, dimensional stability, electricals, and clarity. It should be noted that the very high Izod impact strength values prevail on thicknesses of $1/8''$ but on thicknesses of $1/4''$ or more they drop to about 20% of above values. Even the reduced Izod value is significantly better than a great many engineering plastics. Colorability, hardness, rigidity, and abrasion resistance are other desirable features of PC.

Applications are so numerous that it would be difficult to cite all of them. In the automotive field PC is used for lenses, lamp housings, electrical components, and decorative pieces. Foamed polycarbonate and glass-filled grades are being applied for body parts. FDA grades are used for housewares such as beer mugs, pitchers, returnable bottles for milk, etc.

Electrical and electronic parts, business machine parts, telephone connectors, circuit boards, lighting applications, and other industrial applications use PC owing to the raising of performance standards by public institutions. Glazing for various types of protection, safety shields, and a variety of guards use a grade that has a protective coating for mar resistance and improved chemical resistance.

PC can be processed by injection molding, blow molding, extrusion, and thermoforming.

Power tool housings molded from Lexan resin.

POLYCARBONATE

Category	PROPERTY	1. Lexan-GP	2. Lexan 2014	3. Lexan 940	4. Lexan 191	5. Lexan 500-10% GR	6. Lexan 3412-20% GR	7. Lexan 3413-30% GR	8. Lexan 3414-40% GR	9. Merlon M39	10. Merlon M40	11. Merlon M50
PHYSICAL	Mold shrinkage (in./in.)--in mils	5-7	5-7	5-7	5-7	2-4	3	2.5	2	6-8	6-8	6-8
	Specific gravity	1.2	1.24	1.21	1.19	1.25	1.35	1.43	1.52	1.2	1.2	1.2
	Specific volume (in.^{3}/lb)	23.1	22.3	22.9	23.3	22.2	20.5	19.3	18.2	23.1	23.1	23.1
	Water absorption (%) 24 hr @ 73F (23C)	0.15	0.15	0.15	0.19	0.12	0.16	0.14	0.12	0.12	0.12	0.12
	Water absorption (%) equilibrium, 73F (23C)	0.35	0.32	0.35	0.37	0.31	0.29	0.26	0.23	0.34	0.34	0.34
	Haze (%)	1-2	1-2	1-3	NA	NA	NA	NA	NA	1.0	1.3	1.3
	Transmittance (%)	86-89	85-88	85	NA	NA	NA	NA	NA	88	87	86
MECHANICAL	Tensile strength (psi) yield	9,000	9,000	9,000	8,500	9,600				9,200	9,100	9,000
	Tensile strength (psi) ultimate	9,500	9,500	8,100	9,200	8,000	16,000	19,000	23,000	10,000	10,200	10,500
	Elongation (%) yield	6-8	6-8	6-8	6-8	8-9				8	8	8
	Elongation (%) rupture	110	110	90	119	10-20	4-6	3-5	3-5	115	120	125
	Tensile modulus (x10^5 psi)	3.45	3.45	3.25	3.39	4.5	8.6	12.5	16.8	3.3	3.3	3.3
	Flexural strength (psi)	13,500	13,500	13,200	11,800	15,000	19,000	23,000	27,000	12,500	12,500	12,500
	Flexural modulus (x10^5 psi)	3.4	3.45	3.25	3.10	5.0	8.0	11.0	14.0	3.2	3.2	3.3
	Compressive strength (psi)	12,500	12,500	12,500	10,000	14,000	16,000	18,000	21,000	10,500	10,500	10,500
	Shear strength (psi)	10,000	10,000		8,000	8,500	10,000	10,500	11,000			
	Izod impact strength notched, 1/8 in. thick (ft-lb/in.)	12-16	12	12	14	2	2	2	2.5	15	16	17
	Izod impact strength unnotched, 1/8 in. thick (ft-lb/in.)	No br.	No br.	No br.	No br.	40	19	21	24			
	Tensile impact strength (ft-lb/in.^{2}) S type / L type	225-300	225	250	200	75	30	32	35	275	300	350
	Fatigue strength (psi 2.5x10^6 cycles)	1,000	1,000	1,000	1,000	2,000	5,000	6,000	7,000			
	Rockwell hardness	R118	R120	R118	R118	R124	R122	R120	R119	M62	M62	M62
	Deformation under load (%) @4000 @ 73F (23C)	0.2	0.25		0.22	0.1	0.04	0.03	0.02			
	Deformation under load (%) @4000 @ 158 F	0.3	0.5				0.1	0.08	0.06			
THERMAL	Heat deflection temperature (F) @ 66 psi	280	290	280	280	295	300	305	310			
	Heat deflection temperature (F) @ 264 psi	265-280	270	270	260	288	295	295	295	268	270	271
	Specific heat (Btu/lb/F or Btu/lb/C)	0.3								0.28	0.28	0.28
	Thermal conductivity (Btu/hr/ft^2/F/in.)	1.35	1.35	1.35		1.41	1.47	1.5	1.53	1.34	1.34	1.34
	Coeff. therm. exp. (x10^{-5}) (in./in./F or /C)	3.75	3.75	3.75	3.6	1.79	1.49	1.21	0.93			
	Brittleness temperature (F)	<-200								-150	-150	-150
	Flammability @1/8 in UL Standard 94	94V-2	94V-2	94V-0	94HB	94V-0	94V-0	94V-0	94V-0	94V-2	94V-2	94HB
	Oxygen index	25.0	29.5	35	23.8	32.5	30.5	30	30	14	25	25
ELECTRICAL	Dielectric strength (V/mil)	380	380	425	415	450	490	475	450	>400	>400	>400
	Dielectric constant 60H	3.17	3.17	3.01	3.0	3.1	3.17	3.35	3.53	2.96	2.96	2.96
	Dielectric constant 10^{6}H	2.96	2.96	2.96	2.96	3.05	3.13	3.31	3.48	2.92	2.92	2.92
	Power factor 60H	0.0009	0.0009	0.0009	0.0009	0.0008	0.0009	0.0011	0.0013	0.0006	0.0006	0.0006
	Power factor 10^{6}H	0.010	0.010	0.01	0.0091	0.0075	0.0073	0.007	0.0067	0.009	0.009	0.009

Polycarbonate

Items 1—8 are "Lexan" materials trademarked by General Electric Co. *Items 9—11* "Merlon" numbers are trademarked by Mobay Chemical Co.

Item 1, the "Lexan" G.P. (General Purpose) group, covers three basic grades identified as

#101—High viscosity for parts that require the highest impact strength and for those parts with thick walls (⅛" or more) to minimize sinks and voids.

#141—Medium viscosity for large parts (over 16 square inches) and/or those that are difficult to fill.

#121—Low viscosity for molded parts that have a thin section (0.060" or less) combined with flows of 2" in length or more.

Each one of the basic grades is available in modifications to meet requirements of (a) mold release for parts with little draft, (b) UV protection, and (c) FDA regulations.

Items 2 and 3 are flame-retardant grades for electrical products that must meet today's safety requirements.

Item 4 is applied to thick, (³⁄₁₆" and over) products with the need of highest impact resistance.

Items 5—8 are glass-filled grades as indicated in the headings.

Items 9—11 are "Merlon" basic grades in three viscosities with *Item 9* being the lowest and *Item 11* the highest. These are also available in modifications for mold release, for UV stability, and to meet FDA requirements.

The data and illustrations have been adapted from *Designing with Lexan* by General Electric Co., and *Merlon Polycarbonate—Design Manual* by Mobay Chemical Co. Both of the publications contain additional information that will help in forming a better concept of polycarbonate behavior.

Creep Data for Polycarbonate
(tests were conducted in flexure with support spacing at 4 in.)

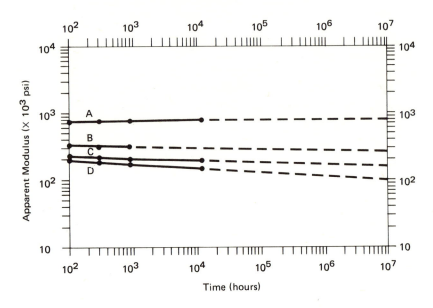

A, Lexan 3414 (40% glass reinforced) at 130°F (54°C) and 4000 psi; B, Lexan 141, 101, and 111 at 73°F (23°C) and 3000 psi; C, Lexan 141, 101, and 111 at 130°F (54°C) and 1500 psi; D, Lexan 141, 101, and 111 at 160°F (71°C) and 1500 psi. Dotted lines indicate extrapolated values.

Stress−strain curve for standard grades of Lexan resins [73°F (23°C)].

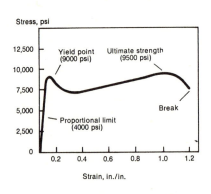

Flexural strength vs temperature for Lexan resins.

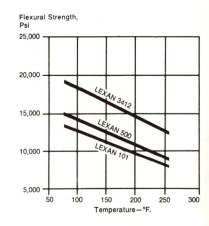

Tensile strength vs temperature for Lexan resins.

Flexural modulus vs temperature for Lexan resins.

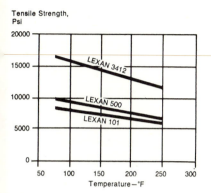

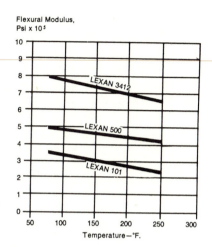

Long-term creep data Lexan resins.

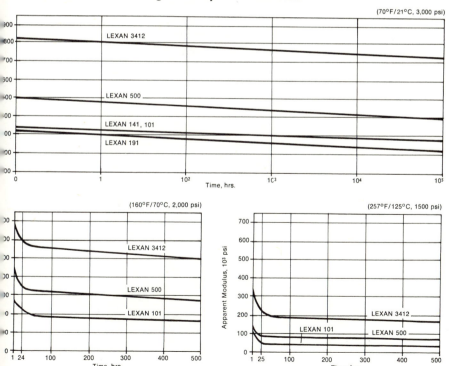

THERMOPLASTIC POLYESTER [POLY(BUTYLENE-TEREPHTHALATE) (PBT)]

PBT is a thermoplastic crystalline resin with a relatively sharp melting point of 440°F (227°C). It is also called a polytetramethylene terephthalate. The PBT resin is the condensation product of 1,4-butanediol and terephthalic acid. The resin displays excellent dimensional stability in environmental extremes; a low coefficient of friction against itself, other plastics, and metals; a low water absorption and relatively high use temperature [250−280°F (121−138°C)]. The electrical characteristics are outstanding and remain so even under conditions of high humidity because of the low water absorption of the material. Colorability is good and, in molded shapes, the appearance is favorable.

The glass-reinforced grades show improved mechanical and creep properties to a degree that they can be considered formidable competitors to thermosets. Flame-retardant grades are also available.

Generally, the thermoplastic polyesters will resist alcohols, ethers, aliphatic hydrocarbons of the type encountered around automotive equipment, and, to some degree, chlorinated hydrocarbons and aqueous salt solutions. The applications of glass-reinforced resin include automotive parts such as guides, gear trains, and parts where a low coefficient of friction is a factor. The material is also used for electrical components in appliances, for hardware items, and for cams and gears for a variety of products. Packaging is an additional outlet for the material. The material is processed predominantly by injection molding, blow molding, and extrusion. It can be painted (oven baked if necessary), printed, cemented, welded, and fabricated.

This valve body with complicated inserts and bosses is made of Celanex.

THERMOPLASTIC POLYESTER

Category	Property	Valox 750 (15% GR)	Valox DR48 (15% GR)	Valox 325 (unreinforced)	Rynite 935 (15% GR)	Rynite 555 (15% GR)	Rynite 530 (30% GR)	Celanex 002 (unreinforced)	Celanex 3311 (30% GR)	Celanex 3210 (20%GR)
PHYSICAL	Mold shrinkage (in./in.)—in mils	5-7	6-8	17-23	4-8	2-7	2-9			
	Specific gravity	1.80	1.53	1.31	1.58	1.80	1.56	1.31	1.65	1.62
	Specific volume (in.^{3}/lb)	15.4	18.7	21.1	17.53	15.39	17.76	21.15	16.7	17.2
	Water absorption (%) 24 hr @ 73F (23C)	0.07	0.07	0.08	0.05	0.04	0.05	0.09	0.07	0.07
	Water absorption, equilibrium, 73F (23C)	0.26	0.30	0.34					0.25	0.25
	Haze (%)									
	Transmittance (%)									
MECHANICAL	Tensile strength (psi) yield									
	Tensile strength (psi) ultimate	12,500	13,000	7,500	14,000	28,500	23,000	8,100	19,500	16,500
	Elongation (%) yield									
	Elongation (%) rupture		5	300	2.2	1.6	2.7	150	1-2	2-3
	Tensile modulus (x10^5 psi)	16.0	8.6	3.44					19.0	12.0
	Flexural strength (psi)	19,500	21,000	12,000	21,500	45,000	33,500	12,500	29,000	24,000
	Flexural modulus (x10^5 psi)	12.0	7.3	3.4	14.0	20.0	13.0	3.73	15.0	11.0
	Compressive strength (psi)		15,000	13,000	20,500	28,500	25,000		22,000	17,000
	Shear strength (psi)		8,000	7,700	7,800	12,000	11,500		9,700	8,000
	Izod impact strength notched, 1/8 in. thick (ft-lb/in.)	0.8	1.0	1.0	1.2	2.3	1.9	0.9	1.4	1.0
	Izod impact strength unnotched, 1/8 in. thick (ft-lb/in.)	6.0	10	No br.					10.0	
	Tensile impact strength (ft-lb/in.^{2}) S type								50	28
	Tensile impact strength (ft-lb/in.^{2}) L type									
	Fatigue strength (psi @10^6 cycles)		4,400	2,850						
	Rockwell hardness		R118	R117			M100	M78	M94	M90
	Deformation under load (%) @ 73F (23C) / @4000 @ 122F						0.4		0.5	1.3
THERMAL	Heat deflection temperature (F) @ 66 psi		410	310				320	410	435
	Heat deflection temperature (F) @ 264 psi	380	360	130	420	440	435	131	395	378
	Specific heat (Btu/lb/F or Btu/lb/C)									0.23
	Thermal conductivity (Btu/hr/ft^2/F/in.)		1.2				2.0		1.63	
	Coeff. therm. exp. (x10^{-5})(in./in./F or /C)	1.6	2.2				1.6		1.3	1.3
	Brittleness temperature (F)									
	Flammability UL Standard 94	94V-0	94V-0	94HB	94HB		94HB	94HB	94V-0	94V-0
	Oxygen index						20		29.5	29.5
ELECTRICAL	Dielectric strength (V/mil) 60H	745	400	590	570		550		530	600
	Dielectric strength (V/mil) 10^{6}H									
	Dielectric constant 60H		3.6	3.1	3.7		3.6			3.7
	Dielectric constant 10^{6}H									
	Power factor 60H		0.02	0.02	0.008		0.005			0.02
	Power factor 10^{6}H				0.010		0.012			
	Volume resistivity (ohm-cm)	2x10^{16}	3.6x10^{16}	4x10^{16}			10^{15}	10^{15}	4x10^{15}	5x10^{15}
	Arc resistance (sec), tungsten electrodes	125	28	184	123		72	72	125	125

Thermoplastic Polyester

Items 1–3 are "Celanex" products, a tradename of Celanese Plastics and Specialties Co.
Items 4–6 are "Rynite" products, trademarked by Du Pont Co.
Items 7–9 are "Valox" products, a tradename of General Electric Co.

When an item is filled with glass, the percentage contained is indicated in the heading as "__% G.R." (glass reinforced).

Basically, all grades listed possess an excellent combination of mechanical, thermal, and electrical properties along with exceptional chemical resistance. The amount and type of reinforcement provide the change in characteristics that is associated with these additives, such as increase in tensile strength, modulus, decrease in impact strength, etc. The choice of grade will depend on matching the product requirement against the data sheet numbers. Each company, along with others, supplies a larger number of grades in addition to those listed.

The General Electric Co. has recently introduced high-impact-strength grades (some as high as 16 ft-lb/in. notched Izod) called Valox VCT or "Valox Chemically Tough." These grades result from new blending technology that permits adding up to 50% of other ingredients, the number of which may be as high as eight. This development can become a source of new and improved grades to meet specific needs.

The data and illustrations have been adapted from suppliers' publications. Additional information can be found in Celanese Bulletin JIA, technical bulletins of Du Pont Co. and *Valox Resin—Design Guide* by General Electric Co.

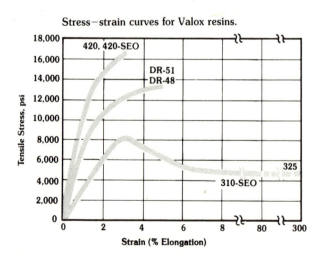

Stress–strain curves for Valox resins.

Flexural creep (Valox 325 resin).

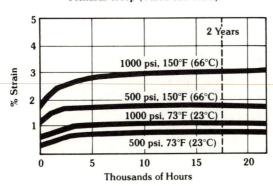

Flexural creep (Valox 420 resin).

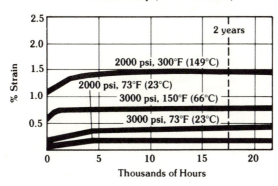

Flexural creep (Valox DR-51 resin).

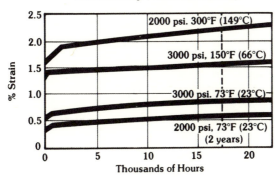

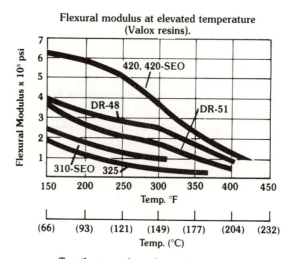

Flexural modulus at elevated temperature
(Valox resins).

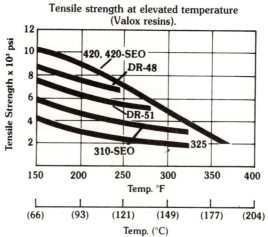

Tensile strength at elevated temperature
(Valox resins).

THERMOPLASTIC COPOLYESTER

This copolyester term is applied to those compounds in which there is more than one glycol and/or more than one dibasic acid. Some of these polyesters can be *amorphous*, some *crystalline*, and others can be either *amorphous or crystalline* depending on processing conditions.

Grades are commercially available as a (1) PCTA polymer of cyclohexanedimethanol (CHDM) and terephthalic acid and another acid that is substituted for an equal displaced part of the terephthalic acid. It is extruded into film that is used as a tough blister packaging material for hardware items; (2) the PETG is a polyethylene-terephthalate-glycol-modified polymer.

The material has good rigidity, toughness, and hardness. When amorphous, it is transparent and virtually colorless.

It is chemically resistant to dilute aqueous solutions of mineral acids, bases, salts, and soaps; it is also resistant to aliphatic hydrocarbons, alcohols, and various oils. Halogenated hydrocarbons, ketones, and aromatic hydrocarbons act as solvents and cause swelling.

Applications for this new material have been some instrument covers, toys, and bottles for packaging. It can be processed by injection and blow molding as well as extrusion.

THERMOPLASTIC COPOLYESTER		GRADE:	1. Kodar PETG 6763
PROPERTY			
PHYSICAL	Mold shrinkage (in./in.)--in mils		
	Specific gravity		1.27
	Specific volume (in.^{3}/lb)		21.81
	Water absorption (%)	24 hr @ 73F (23C)	
		equilibrium, 73F (23C)	
	Haze (%)		
	Transmittance (%)		
MECHANICAL	Tensile strength (psi)	yield	7,100
		ultimate	4,000
	Elongation (%)	yield	
		rupture	180
	Tensile modulus (x10^5 psi)		
	Flexural strength (psi)		10,000
	Flexural modulus (x10^5 psi)		2.9
	Compressive strength (psi)		
	Shear strength (psi)		
	Izod impact strength notched, 1/8 in.thick		1.7
	(ft-lb/in.) unnotched, 1/8 in.thick		No br.
	Tensile impact strength (ft-lb/in.^{2})	S type	
		L type	
	Fatigue strength (psi @ ___ cycles)		
	Rockwell hardness		R105
	Deformation under load (%)	@ ___ @ 73F (23C)	
		@ ___ @ ___ F	
THERMAL	Heat deflection temperature (F)	@ 66 psi	158
		@ 264 psi	145
	Specific heat (Btu/lb/F or Btu/lb/C)		
	Thermal conductivity (Btu/hr/ft^2/F/in.)		
	Coeff. therm. exp.(x10^{-5}) (in./in./F or /C)		
	Brittleness temperature (F)		
	Flammability	UL Standard 94	
		Oxygen index	
ELECTRICAL	Dielectric strength (V/mil)		400
	Dielectric constant	60H	
		10^{6}H	3.2
	Power factor	60H	
		10^{6}H	0.3
	Volume resistivity (ohm-cm)		7x10^{13}
	Arc resistance (sec), tungsten electrodes		150

POLYESTER, AROMATIC (POLYARYLATE)

This is one of the newer thermoplastic materials. It is an aromatic polyester of phthalic acids and bisphenols, and its structure is amorphous. The following are its outstanding characteristics: high modulus, good flexural recovery, good UV stability, good flammabiltiy property, high heat deflection temperature, good electrical properties, high toughness with low notch sensitivity, excellent creep resistance, exceptional thermal resistance. It is transparent with a light golden color.

The potential applications are for outdoor products, safety devices, transportation, electronic and electrical components, snapfit assemblies, lighting, glazing, and solar energy components. The material can be processed on screw injection machines and extrusion facilities.

POLYESTER, AROMATIC (POLYARYLATE) PROPERTY	GRADE	1. Ardel D-100
PHYSICAL		
Mold shrinkage (in./in.)--in mils		9
Specific gravity		1.2
Specific volume (in.^{3}/lb)		22.89
Water absorption (%)	24 hr @ 73F (23C)	0.27
	equilibrium, 73F (23C)	0.71
Haze (%)		
Transmittance (%)		
MECHANICAL		
Tensile strength (psi)	yield	9,500
	ultimate	
Elongation (%)	yield	8
	rupture	50
Tensile modulus (x10^5 psi)		2.9
Flexural strength (psi)		11,000
Flexural modulus (x10^5 psi)		3.1
Compressive strength (psi)		
Shear strength (psi)		
Izod impact strength (ft-lb/in.)	notched, 1/8 in.thick	4.2
	unnotched, 1/8 in.thick	
Tensile impact strength (ft-lb/in.^{2})	S type	
	L type	
Fatigue strength (psi @ ___ cycles)		
Rockwell hardness		
Deformation under load (%) @ ___	@ 73F (23C)	
	@ ___ @ ___ F	
THERMAL		
Heat deflection temperature (F)	@ 66 psi	
	@ 264 psi	345
Specific heat (Btu/lb/F or Btu/lb/C)		
Thermal conductivity (Btu/hr/ft^2/F/in.)		
Coeff. therm. exp.(x10^{-5}) (in./in./F or /C)		
Brittleness temperature (F)		
Flammability	UL Standard 94	94V-0
	Oxygen index	34
ELECTRICAL		
Dielectric strength (V/mil)		400
Dielectric constant	60H	2.73
	10^{6}H	2.60
Power factor	60H	0.0008
	10^{6}H	0.02
Volume resistivity (ohm-cm)		
Arc resistance (sec), tungsten electrodes		125

POLYETHERIMIDE (ULTEM)

Ultem is a development and a tradename of the General Electric Co. It is an amorphous thermoplastic. The outstanding characteristics of the material are:

Thermal Properties. The material is rated by UL for continuous use at 338°F (170°C), and, with a high heat deflection temperature of 392°F (200°C) at 264 psi, it is capable of retaining relatively high physical properties at elevated temperatures.

The thermal coefficient of expansion is low, making it easier to design for products where plastic works in conjunction with metals.

The flame resistance is inherent to the material, and the oxygen index is higher than any of the other engineering thermoplastics.

Mechanical Properties. The tensile and flexural strengths, the modulus, and the stress–strain relationship are all well above most of the engineering thermoplastics. For example, at 375°F (191°C) Ultem unfilled resin retains a tensile strength of 6000 psi; the proportional limit of Ultem 1000 is 8000 psi. At room temperature and at stress levels of 5000 psi, the slope of the flexural creep line (apparent modulus vs time) is small, pointing to a low percentage change of apparent modulus over extended periods of time. At elevated temperatures and stress levels under consideration for specific designs, the creep behavior should be investigated.

Other interesting properties are high dielectric strength; stable dielectric constant and dissipation factor over a wide range of temperature and frequencies; broad chemical resistance and transparency.

The data sheets and pertinent curves will show the favorable aspects of the material in relation to other characteristics.

The important consideration in evaluating all the desirable properties of Ultem is that they can be attained with ease of processing, i.e., conversion of the raw material into finished products that are comparable to those of the popular engineering thermoplastics.

Polyetherimide

A more-detailed listing and analysis of the properties, design, processing, and secondary operations can be found in *Ultem* by General Electric Co.

The data and illustrations have been adapted from that publication.

POLYETHERIMIDE

PROPERTY	1. Ultem 1000 (unfilled)	2. Ultem 2100 (10% GR)	3. Ultem 2200 (20% GR)	4. Ultem 2300 (30% GR)
PHYSICAL				
Mold shrinkage (in./in.)--in mils	5-7	4	2-3	2
Specific gravity	1.27	1.34	1.42	1.51
Specific volume (in.^{3}/lb)	21.81	20.67	19.51	18.34
Water absorption (%) 24 hr @ 73F (23C)	0.25	0.28	0.26	0.18
equilibrium, 73F (23C)	1.25	1.0	1.0	0.9
Haze (%)				
Transmittance (%)				
MECHANICAL				
Tensile strength (psi) yield	15,200	16,600	20,100	24,500
ultimate				
Elongation (%) yield	7-8	5		
rupture	60	6	3	3
Tensile modulus (x10^5 psi)	4.3	6.5	10.0	13.0
Flexural strength (psi)	21,000	28,000	30,000	33,000
Flexural modulus (x10^5 psi)	4.8	6.5	9.0	12.0
Compressive strength (psi)	20,300	22,500	24,500	23,500
Shear strength (psi)	15,000	13,000	13,500	14,000
Izod impact strength notched, 1/8 in. thick	1.0	1.1	1.6	2.0
(ft-lb/in.) unnotched, 1/8 in. thick	25	9.0	9.0	8.0
Tensile impact strength (ft-lb/in.^{2}) S type				
L type				
Fatigue strength (psi @ ___ cycles)				
Rockwell hardness	M109	M114	M118	M125
Deformation under load (%) @ ___ @ 73F (23C)				
@ ___ @ ___ F				
THERMAL				
Heat deflection temperature (F) @ 66 psi	410	410	410	414
@ 264 psi	392	405	408	410
Specific heat (Btu/lb/F or Btu/lb/C)				
Thermal conductivity (Btu/hr/ft^2/F/in.)				
Coeff. therm. exp.(x10^{-5}) (in./in./F or /C)	3.4	1.8	1.4	1.1
Brittleness temperature (F)				
Flammability UL Standard 94	94V-0	94V-0	94V-0	94V-0
Oxygen index	47		50	
ELECTRICAL				
Dielectric strength (V/mil)	831			769
Dielectric constant 60H				
10^{6}H	3.15	3.5	3.5	
Power factor 60H				
10^{6}H	0.0013	0.0014	0.0015	0.0015
Volume resistivity (ohm-cm)				
Arc resistance (sec), tungsten electrodes	128			85

Stress–strain curves in tension at 73°F (23°C).

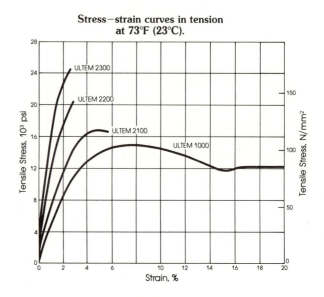

Stress–strain curves in tension at 200°F (93°C).

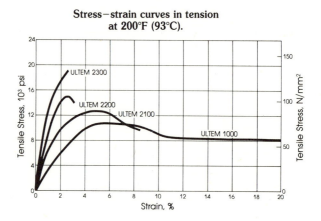

Stress–strain curves in tension at 350°F (177°C).

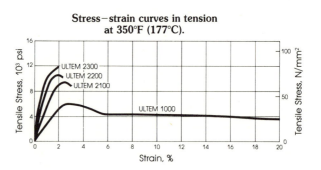

Tensile strength of Ultem™ as a function of temperature.

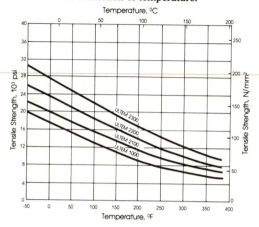

Flexural modulus as a function of temperature.

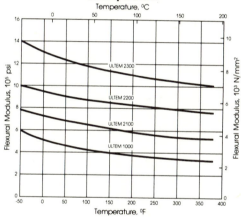

Flexural creep at 73°F (23°C).

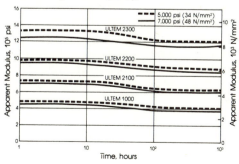

POLYETHYLENE (PE)

Polyethylene is a thermoplastic and a member of the polyolefin family. It is prepared by the polymerization of ethylene as the sole monomer. In the plastic industry it is called the "poor man's" Teflon because of its outstanding resistance to all types of chemicals. Its chemical structure is simple in comparison with other polymers; on the other hand, it can be produced in innumerable grades to meet specific needs, and it would be difficult to define the properties covering them all.

Generally speaking the material has excellent electrical insulating properties, it is inert to many chemicals, it retains its flexibility at low temperature [$-100°F$ ($-73°C$)], it can be formulated to be soft or hard and rigid, tough, and flexible for impact resistance. With the inclusion of additives, reinforcements, and other copolymers the range of properties can be varied to suit hundreds of different applications. One important way to characterize the polyethylenes is by density (specific gravity). With increased density, the following will increase: the tensile strength, rigidity, hardness, resistance to heat and creep, chemical resistance, surface gloss, and barrier properties.

The densities range from 0.91 to 0.96 and over. In addition to the designation by density there is one of polymerization processing. LDPE—*low*-density polyethylene—is produced by the *high*-pressure process in which the molecules show a considerable amount of branching. These branches do not permit uniform arrangement of molecular chains during solidification with the result that the product ends up with a lower density and a predominant amorphous structure. Copolymers of ethylene with other comonomers can also be produced by this process.

HDPE—*high*-density polyethylene—also called linear low-density polyethylene is produced by the *low*-pressure process in which little branching exists and the more-linear molecules can arrange in a more orderly and denser manner with the result that the solid material is stiffer and harder, denser, and more crystalline and displays a higher shrinkage. Another designation is the melt index, which is an indication of the flow properties of a compound. It establishes the flow of material over a set period of time, through an orifice with a predetermined pressure and material temperature.

LDPE is used for packing film, houseware items, blow-molded containers, wire and cable insulation, and extruded products.

HDPE is used for blow-molded containers for milk and household chemicals, household products, large rotomolded parts, cable coatings, and film.

As indicated before there are copolymers with other comonomers and various additives and reinforcements that extend the use of the material into thousands of applications.

Polyethylenes can be processed by every known method used for thermoplastics such as injection-, blow- and rotomolding, extrusion, thermoforming, and various fabrication means.

POLYETHYLENE

PROPERTY	IP-FF 1008 (GR)	IP-FF 1006 (GR)	IP-FF 1004 (GR)	arlex R880	arlex 53120	moco 29081	moco 2	lathon	lathon
PHYSICAL									
Mold shrinkage (in./in.)--in mils	2.5	3.0	3.5					<20	<20
Density (g/cm^3)	1.28	1.17	1.10	0.955	0.953	0.952	0.952	0.94	0.96
Specific volume (in.^{3}/lb)	21.6	23.7	25.2	29.01	29.07	29.10	29.10	29.47	28.85
Water absorption (%) 24 hr @ 73F (23C)	0.020	0.015	0.010						
Water absorption (%) equilibrium, 73F (23C)	0.30	0.20	0.10						
Haze (%)									
Transmittance (%)									
MECHANICAL									
Tensile strength (psi) yield	11,500	10,000	8,000	4,200	3,800	4,000	4,000	3,300	4,200
Tensile strength (psi) ultimate									
Elongation (%) yield	2-3	2-3	2-3			8.5	9		
Elongation (%) rupture				600	>650	300	1,300	800	900
Tensile modulus ($\times 10^5$ psi)				2.0	1.9	1.55	1.55	1.4	2.2
Flexural strength (psi)	14,000	11,500	10,000						
Flexural modulus ($\times 10^5$ psi)	11.0	9.0	6.0						
Compressive strength (psi)	4,900	4,400	4,000						
Shear strength (psi)									
Izod impact strength notched, 1/8 in.thick (ft-lb/in.)	1.3	1.1	1.0			0.8	1.2	1.5	1.8
Izod impact strength unnotched, 1/8 in.thick	11-12	8-9	4-5						
Tensile impact strength (ft-lb/in.^{2}) S type	31	28	25			26	42	145	130
Tensile impact strength (ft-lb/in.^{2}) L type									
Fatigue strength (psi @ ___ cycles)									
Rockwell hardness (D-Durometer)	R90	R85	R80			D66	D67	D59	D65
Deformation under load (%) @ __ @ __ @ __ F 73F (23C)									
THERMAL									
Heat deflection temperature (F) @ 66 psi	270	270	265					140	160
Heat deflection temperature (F) @ 264 psi	260	260	250						
Specific heat (Btu/lb/F or Btu/lb/C)									
Thermal conductivity (Btu/hr/ft^2/F/in.)									
Coeff. therm. exp.($\times 10^{-5}$)(in./in./F or /C)	2.4	2.7	3.0						
Brittleness temperature (F)								<-105	<-105
Flammability UL Standard 94									
Oxygen index									
ELECTRICAL									
Dielectric strength (V/mil)									
Dielectric constant 60H									
Dielectric constant 10^6H									
Power factor 60H									
Power factor 10^6H									
Volume resistivity (ohm-cm)									
Arc resistance (sec), tungsten electrodes									

Polyethylene

Polyethylene is considered a utility resin and, for this reason, the available information on properties is limited to that of interest to its users. When the resin is reinforced with glass fibers, it gains in properties that are of interest to designers.

Items 1 and 2 are "Alathon," a trademark of the Du Pont Co.

Item 1 has high stiffness, good dimensional stability, and resistance to warpage.

Item 2 provides toughness and stress crack resistance, while retaining the desirable properties of *Item 1* to a large degree.

Item 3 is an Amoco grade, suggested for such applications as hard hats and beverage cases.

Item 4 is also made by Amoco and is offered for thin-wall containers and overcaps.

Items 5 and 6 are "Marlex," a trademark of Phillips Chemical Co. *Item 5* is suggested for application to crates and industrial parts. *Item 6* is being applied to milk cases, tote boxes, and automotive components.

Items 7−9 are glass-reinforced products of LNP and the potential uses are automotive fender liners, blower and fan casings, chemical processing pipe fittings, and pump components.

The number of grades from each supplier, as well as the number of suppliers, is quite large.

ULTRA-HIGH-MOLECULAR-WEIGHT POLYETHYLENE (UHMWPE)

Ultra-high-molecular-weight polyethylene (UHMWPE) is a high-density polyethylene with a molecular weight of $2-6$ million (10^6). The resins supplied for pipe and called high-molecular-weight materials have by comparison a molecular weight of $300,000-500,000$.

UHMWPE possesses many desirable properties such as (1) the highest abrasion resistance of any thermoplastic, (2) low coefficient of friction, (3) self-lubrication, (4) nonstick surfaces, (5) exceptional impact resistance even at cryogenic temperatures, (6) good chemical resistance, (7) good stress-cracking resistance, (8) good fatigue resistance, (9) sound dampening properties, and (10) FDA plus USDA sanction for use in food, meat, and poultry applications.

These properties and others can be enhanced by the addition of graphite fibers, glass fibers or glass beads, talc, or other additives, because the desired improvements are usually in the direction of rigidity, resistance to deformation and creep, and increased heat deflection temperature at little or no sacrifice of inherent material properties.

The principal applications are for processing machinery and materials-handling facilities where the machinery components are in contact with chemicals, food, beverages, or medicine or for cases where exposure is to highly abrasive action of materials being handled. An example are chutes where the use of a material with abrasion resistance and a low coefficient of friction is a necessity. There are numerous applications where abrasion resistance is a problem, and, consequently, this material is considered in all such cases. One of the more interesting uses is in orthopedic hip, knee, or other joints for humans.

UHMWPE is supplied in a fine powder form and is shaped by a modified compression-molding process. Sheet can be continuously formed on special machinery. An injection machine was developed in Japan that will mold parts to needed configuration. Most of the parts presently are machined from blocks, sheet, rods, and tubing with woodworking-like tools and machinery.

ULTRA-HIGH-MOLECULAR-WEIGHT POLYETHYLENE	GRADE 1. Hostalen GUR
PROPERTY	
PHYSICAL	
Mold shrinkage (in./in.)--in mils	40
Specific gravity	0.94
Specific volume (in.^{3}/lb)	29.47
Water absorption (%) 24 hr @ 73F (23C)	0
equilibrium, 73F (23C)	
Haze (%)	
Transmittance (%)	
MECHANICAL	
Tensile strength (psi) yield	3,100
ultimate	6,200
Elongation (%) yield	
rupture	600
Tensile modulus (x10^5 psi)	1.5
Flexural strength (psi)	
Flexural modulus (x10^5 psi)	
Compressive strength (psi)	
Shear strength (psi)	
Izod impact strength notched, 1/8 in.thick	30
(ft-lb/in.) unnotched, 1/8 in.thick	
Tensile impact strength (ft-lb/in.^{2}) S type	
L type	
Fatigue strength (psi @ ___ cycles)	
Rockwell hardness	
Deformation under load (%) @ ___ @ 73F (23C)	
@ ___ @ ___ F	
THERMAL	
Heat deflection temperature (F) @ 66 psi	155-180
@ 264 psi	110-120
Specific heat (Btu/lb/F or Btu/lb/C)	
Thermal conductivity (Btu/hr/ft^2/F/in.)	
Coeff. therm. exp.(x10^{-5}) (in./in./F or /C)	
Brittleness temperature (F)	
Flammability UL Standard 94	
Oxygen index	
ELECTRICAL	
Dielectric strength (V/mil)	900
Dielectric constant 60H	2.3
10^{6}H	
Power factor 60H	
10^{6}H	
Volume resistivity (ohm-cm)	10^{18}
Arc resistance (sec), tungsten electrodes	

POLYIMIDE

Polyimide—Thermoplastic

Thermoplastic polyimide has rings of four carbon atoms tightly bound together. The material is thermoplastic in the sense that it has a glass transition point when it changes into a rubbery state at 590°F (310°C). It does not have a melting point that would permit conventional injection molding.

This material's main characteristic is the retention of excellent properties at temperatures of 500°F (260°C) and over for prolonged periods of time (thousands of hours). It is reinforced with glass fabrics, quartz fibers, boron fibers, graphite and glass fibers, as well as with chopped glass, graphite, molybdenum disulfide, and PTFE.

The material has inherently good flammability characteristics. When exposed to heat up to 900°F (482°C) for short periods of time, it will perform in a satisfactory manner. Its creep resistance is outstanding, and other mechanical properties at temperatures up to 480°F (249°C) will only drop to 70% of those values that are indicated at tests of 73°F (23°C). The coefficient of thermal expansion comes close to that of metals. It is not affected by nuclear radiation, dilute acids, aromatic and aliphatic hydrocarbons, esters, ethers, alcohols, kerosene, and hydraulic fluids. It is attacked by dilute alkalies and inorganic acids. The material is available as filled and unfilled molding compound. An injection-molding compound consisting of 60% polyimide and 40% polyphenylene sulfide is available for use on standard injection machines. The applications are for piston rings, valves seats, bearings, and seals that will perform satisfactorily at 500°F (260°C). Other components for electrical and mechanical devices around hot engines also utilize this material. The material is processed by compression molding by heating the mold above the glass transition temperature just enough to permit good flow. Film can be cast from solution with conventional casting machines. Fibers can be solution-spun for high-temperature-resistance fabrics.

Polyimide—Thermoset

The thermoset grade of polyimide is very similar in every respect to the thermoplastic one, except that the latter has a higher elongation and displays greater toughness.

The thermoset grade is easier to mold; therefore, it lends itself to forming shapes for engineering applications. Thus we find applications such as aircraft engine parts, high-performance automotive parts, gears, coil bobbins, printed circuit boards, insulation for high-performance electric motors, and similar applications where high heat resistance over prolonged periods of time is a factor.

Processing can be similar to powder metallurgy (a combination of compression and sintering) or compression transfer and injection molding.

POLYIMIDE

PROPERTY		1. Thermoplastic unfilled	2. Thermoplastic 40% graphite fiber filled	3. Thermosetting 50% GR
PHYSICAL Mold shrinkage (in./in.)--in mils				2
Specific gravity		1.4	1.65	1.65
Specific volume (in.^{3}/lb)		19.79	16.79	16.79
Water absorption (%) 24 hr @ 73F (23C)		0.24	0.14	0.7
equilibrium, 73F (23C)		1.2	0.6	
Haze (%)				
Transmittance (%)				
MECHANICAL Tensile strength (psi) yield		12,500		
ultimate		14,000	7,600	6,400
Elongation (%) yield				
rupture		9	3	
Tensile modulus (x10^5 psi)		3.0		
Flexural strength (psi)		20,000	14,000	21,000
Flexural modulus (x10^5 psi)		5.0	7.0	20.0
Compressive strength (psi)		30,000	16,000	34,000
Shear strength (psi)				
Izod impact strength notched, 1/8 in.thick		1.5	0.7	5.6
(ft-lb/in.) unnotched, 1/8 in.thick				
Tensile impact strength (ft-lb/in.^{2}) S type				
L type				
Fatigue strength (psi @ ___ cycles)				
Rockwell hardness				M118
Deformation under load (%) @ ___ @ 73F (23C)				
@ ___ @ ___ F				
THERMAL Heat deflection temperature (F) @ 66 psi				
@ 264 psi		600	680	660
Specific heat (Btu/lb/F or Btu/lb/C)				
Thermal conductivity (Btu/hr/ft^2/F/in.)				
Coeff. therm. exp.(x10^{-5}) (in./in./F or /C)				
Brittleness temperature (F)				
Flammability UL Standard 94				
Oxygen index				
ELECTRICAL Dielectric strength (V/mil)		560		450
Dielectric constant 60H				
10^{6}H				
Power factor 60H				
10^{6}H				
Volume resistivity (ohm-cm)				
Arc resistance (sec), tungsten electrodes				

POLYMETHYLPENTENE (TPX)

Polymethylpentene is a thermoplastic and one of the lightest members of the polyolefin family.

The distinguishing properties of the commercial compounds are good transparency (90% light transmission), good electrical properties, good chemical resistance, useful mechanical properties at 400°F (204°C), and low specific gravity of 0.83. The resistance to chemicals is similar to other olefins. It is attacked by strong oxidizing agents, and, when exposed to some light hydrocarbons and chlorinated solvents, it will swell and lose some of the mechanical properties.

It is used for laboratory ware, for automotive, appliance, electrical parts, and for packaging containers.

It can be injection-blow-molded and extruded.

POLYMETHYLPENTENE PROPERTY		GRADE	1. Unfilled resin
PHYSICAL	Mold shrinkage (in./in.)--in mils		15-30
	Specific gravity		0.83
	Specific volume (in.^{3}/lb)		33.37
	Water absorption (%) 24 hr @ 73F (23C)		0.01
	equilibrium, 73F (23C)		
	Haze (%)		
	Transmittance (%)		
MECHANICAL	Tensile strength (psi) yield		
	ultimate		3,500
	Elongation (%) yield		
	rupture		50
	Tensile modulus (x10^5 psi)		1.6
	Flexural strength (psi)		4,000
	Flexural modulus (x10^5 psi)		1.1
	Compressive strength (psi)		5,000
	Shear strength (psi)		
	Izod impact strength notched,1/8 in.thick		0.5
	(ft-lb/in.) unnotched,1/8 in.thick		
	Tensile impact strength (ft-lb/in.^{2}) S type		
	L type		
	Fatigue strength (psi @ ___ cycles)		
	Rockwell hardness		
	Deformation under load (%) @ ___ @ 73F (23C)		
	@ ___ @ ___ F		
THERMAL	Heat deflection temperature (F) @ 66 psi		212
	@ 264 psi		106
	Specific heat (Btu/lb/F or Btu/lb/C)		
	Thermal conductivity (Btu/hr/ft^2/F/in.)		
	Coeff. therm. exp.(x10^{-5}) (in./in./F or /C)		
	Brittleness temperature (F)		
	Flammability UL Standard 94		
	Oxygen index		
ELECTRICAL	Dielectric strength (V/mil)		
	Dielectric constant 60H		
	10^{6}H		
	Power factor 60H		
	10^{6}H		
	Volume resistivity (ohm-cm)		
	Arc resistance (sec), tungsten electrodes		

POLYPHENYLENE SULFIDE (PPS) (TRADENAME RYTON)

PPS is an aromatic thermoplastic with a high degree of crystallinity in structure.

PPS has outstanding heat stability, good inherent flammability property, and good chemical resistance. It is rigid and retains good mechanical properties at elevated temperatures [500°F (260°C)]. The improvement in properties when reinforced with fiberglass is above the average of other thermoplastics, which is owing to the exceptional ability of the resin to wet the glass fibers. This can be observed as a tripling of the Izod impact strength for the 40% glass-filled grade.

Another grade filled with glass and mineral filler is formulated for arc resistance and low track rate in demanding electrical uses.

PPS is inert to a wide variety of organic solvents, aqueous inorganic salts, as well as acids and bases and, thus, can be exposed to chemical environments without ill effects upon its performance. The material also has good colorability.

The applications are in processing machinery components, for all types of pumps where the requirement is good chemical and heat resistance. Electronic and computer parts take advantage of its good electrical properties at elevated temperatures. Hair dryers that are handheld and require heat resistance coupled with good appearance use PPS for housings. Some automotive parts use PPS because of its resistance to corrosive effects of exhaust gases, gasoline, and ethylene glycol.

Processing is done by injection molding and extrusion.

Polyphenylene Sulfide

Items 1−6 are "Ryton," a trademark of Phillips Chemical Co., *Items 7 and 8* are products of LNP Corp. *Items 2 and 3* are 40% glass fiber filled. *Items 4 and 5* are glass fiber and mineral filled. *Item 6* has a higher percentage of glass fiber and mineral filling than the preceding ones.

Additional information on the characteristics of the material can be found in *"Ryton" 100—Properties—Processing* and *"Ryton" TSM-266—* *Physical, Chemical, Thermal and Electrical Properties* by Phillips Chemical Co.

The data and illustrations have been adapted from the publications of the suppliers.

POLYPHENYLENE SULFIDE

	PROPERTY	Ryton (unfilled)	Ryton R-4 (as molded)	Ryton R-4 (annealed)	Ryton R-8 (as molded)	Ryton R-8 (annealed)	Ryton R-10 (annealed)	LNPOF 1006 (30% GR)	LNPOF 1008 (40% GR)
PHYSICAL	Mold shrinkage (in./in.)--in mils	10						2	2
	Specific gravity	1.34	1.6	1.6	1.69	1.69	1.99	1.56	1.65
	Specific volume (in.^{3}/lb)	20.67	17.28	17.28	16.43	16.43	13.94	17.75	16.79
	Water absorption (%) 24 hr @ 73F (23C)	0.05	<0.05	<0.05	0.03	0.03	0.06	0.04	0.02
	equilibrium, 73F (23C)								
	Haze (%)								
	Transmittance (%)								
MECHANICAL	Tensile strength (psi) yield								
	ultimate	10,800	19,000	16,200	13,200	9,000	11,500	20,000	23,000
	Elongation (%) yield								
	rupture		1.25	0.68	0.64	0.40	0.06	3-4	3-4
	Tensile modulus ($\times 10^5$ psi)								
	Flexural strength (psi)	14,000	26,800	24,700	21,000	13,000	18,500	29,000	32,000
	Flexural modulus ($\times 10^5$ psi)	6.0	17.0	17.0	20.0	21.0	18.0	16.0	18.0
	Compressive strength (psi)		21,000	21,000	16,000	16,000	18,500		
	Shear strength (psi)								
	Izod impact strength notched, 1/8 in.thick (ft-lb/in.)	0.3	1.4	1.2	0.59	0.57	1.0	1.4	1.5
	unnotched, 1/8 in.thick	3-4	5.9	2.7	2.1	1.5	2.5	8-9	11-12
	Tensile impact strength (ft-lb/in.^{2}) S type / L type								
	Fatigue strength (psi @ cycles)								
	Rockwell hardness		R123	R123	R121	R121	R120		
	Deformation under load (%) @ 73F (23C) @ °F								
THERMAL	Heat deflection temperature (F) @ 66 psi								
	@ 264 psi	280	470	>500	471	>500	>500	500	505
	Specific heat (Btu/lb/F or Btu/lb/C)		0.25	0.25					
	Thermal conductivity (Btu/hr/ft^2/F/in.)	2.0	2.0	2.0	1.6	1.6		2.8	3.1
	Coeff. therm. exp.($\times 10^{-5}$) (in./in./F or /C)	3	2.2	2.2	1.6			1.3	1.5
	Brittleness temperature (F)								
	Flammability UL Standard 94	94V-0	94V-0/5	94V-0/5	94V-0/5	94V-0/5	94V-0	94V-0	94V-0
	Oxygen index		46.5	46.5	53	53			
ELECTRICAL	Dielectric strength (V/mil)	380	450	450	340	340	400	375	375
	Dielectric constant 60H	3.1	3.8	3.8	4.3	4.3	4.8-6.1	3.6	3.8
	10^{6}H								
	Power factor 60H	0.0009	0.0014	0.0014	0.016	0.016	.01-.02	0.0014	0.0014
	10^{6}H								
	Volume resistivity (ohm-cm)	4.5×10^{16}	4.5×10^{16}	4.5×10^{16}	2×10^{15}	2×10^{15}	2×10^{15}	4.7×10^{16}	5.0×10^{16}
	Arc resistance (sec), tungsten electrodes		180	180	235	235	160		

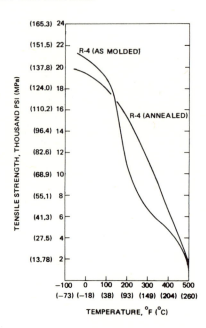

Ryton R-4 tensile strength vs temperature.

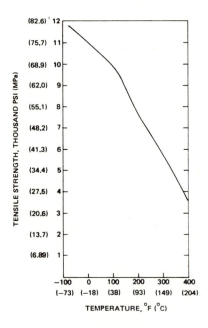

Ryton R-8 tensile strength vs temperature
(as molded).

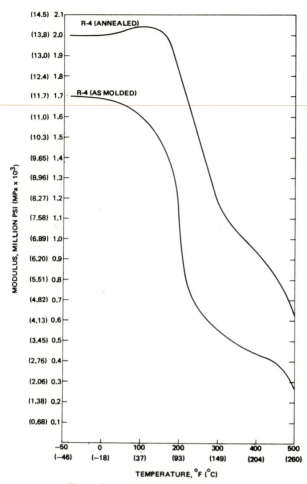

Flexural modulus vs temperature for Ryton R-4.

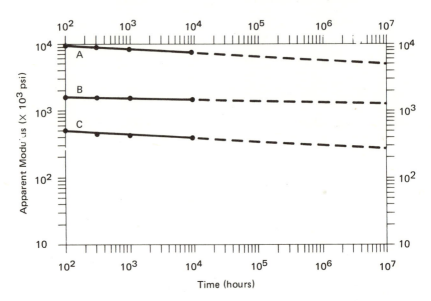

Creep Data for Polyphenylsulfone (Ryton–R4 annealed)
(tests were conducted in flexure with support spacing at 4 in.)

A, 73°F (23°C) at 1000 psi; B, 150°F (66°C) and 5000 psi; C, 250°F (121°C) at 5000 psi.
Dotted lines indicate extrapolated values.

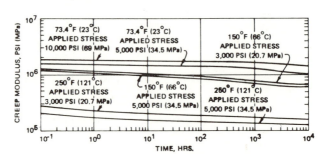

Creep (apparent) modulus of unannealed Ryton R-4.

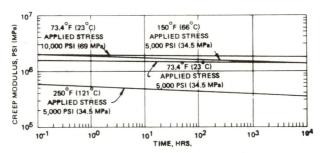

Creep (apparent) modulus annealed Ryton R-4.

POLYPROPYLENE (PP)

PP is a crystalline thermoplastic material and is a member of the polyolefin family. It is prepared by the polymerization of propylene as the sole monomer. PP is characterized as having good chemical resistance, rigidity, electrical properties, moderate heat resistance, and unusual fatigue resistance in flexure. Most of the properties are related to melt index (molecular-weight flow rate), degree of crystallinity, molecular orientation, and proportion of isotactic to atactic structure.

At higher flow rates, the tensile and flexural strengths and heat deflection temperature will increase, whereas elongation and impact strengths will decrease. At boiling water heat [212°F (100°C)] good mechanical properties are maintained. The material is, however, subject to oxidative degradation at elevated temperatures. This deficiency can be overcome with appropriate antioxidant stabilizers as an additive to the base polymer.

PP resists solutions of inorganic salts, mineral acids, and bases. Surface swelling occurs when exposed to aromatic and chlorinated hydrocarbons. Most of the organic compounds have no ill effects on this polymer.

A variety of grades is available for special uses by varying flow properties, additives, fillers, reinforcing agents, and elastomeric compounds and by copolymerization with ethylene.

The applications are extensive, particularly in the automotive, housewares, appliance, medical, and packaging fields. A large variety of parts are made in each industry owing to the low cost of the material and to the ability to compound it to suit requirements. One of the largest single uses is storage battery cases. Polypropylene fiber is used for cordage, industrial bagging, and carpeting.

It can be processed by injection and blow molding by extrusion, hot forming, and any other process applicable to thermoplastics.

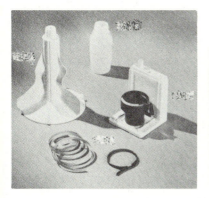

Typical parts made with polypropylene.

POLYPROPYLENE

	PROPERTY	GRADE →	1. Amoco 1012	2. Amoco 1046	3. Amoco 2226	4. LNP MFX 1004 HS	5. LNP MFX 1008 HS	6. Tenite 4240	7. Tenite 5021	8. Tenite 4P31	9. Tenite 4K31
PHYSICAL	Mold shrinkage (in./in.)--in mils		15-20	15-20	10-17	4	3				
	Specific gravity		0.9	0.9	1.03	1.04	1.22	0.902	0.899	0.90	1.04
	Specific volume (in.^{3}/lb)		30.8	30.8	26.89	26.6	22.7	30.7	30.81	30.8	26.6
	Water absorption (%) 24 hr @ 73F (23C)		<0.01	<0.01	0.01	0.01	0.05				
	Water absorption (%) equilibrium, 73F (23C)					0.02	0.09				
	Haze (%)										
	Transmittance (%)										
MECHANICAL	Tensile strength (psi) yield		5,400	5,400	5,200	11,500	15,000	4,800	3,800	4,000	4,800
	Tensile strength (psi) ultimate										
	Elongation (%) yield										
	Elongation (%) rupture		40	120	30	4-5	3-4				
	Tensile modulus (x10^5 psi)										
	Flexural strength (psi)					14,000	22,000				
	Flexural modulus (x10^5 psi)		2.1	2.2	3.75	6.0	10.0	1.7	1.2	1.5	3.1
	Compressive strength (psi)										
	Shear strength (psi)										
	Izod impact strength notched, 1/8 in.thick (ft-lb/in.)		1.4	0.5	0.45	1.8	2.0	0.6	1.0	2.5	0.7
	Izod impact strength unnotched, 1/8 in.thick (ft-lb/in.)					10-12	11-13	>25	No br.	>25	15
	Tensile impact strength (ft-lb/in.^{2}) S type / L type		65	35	12						
	Fatigue strength (psi @ ___ cycles)										
	Rockwell hardness (D-Durometer)		D75	D76	D77	R111	R111	R90	R70	R77	R88
	Deformation under load (%) @ 73F (23C) @ ___ psi @ ___ F										
THERMAL	Heat deflection temperature (F) @ 66 psi		215	230	270	320	330				
	Heat deflection temperature (F) @ 264 psi					305	310	127	122	126	151
	Specific heat (Btu/lb/F or Btu/lb/C)					0.44	0.38				
	Thermal conductivity (Btu/hr/ft^2/F/in.)					2.1	2.55				
	Coeff. therm. exp.(x10^{-5})(in./in./F or /C)					2.5	1.5				
	Brittleness temperature (F)										
	Flammability UL Standard 94		94HB	94HB	94HB	94HB	94HB				
	Oxygen index										
ELECTRICAL	Dielectric strength (V/mil) 60H / 10^{6}H										
	Dielectric constant 60H / 10^{6}H										
	Power factor										

POLYPROPYLENE (continued)

PROPERTY		Pro-fax 3	Pro-fax 3	Pro-fax 4-2	Pro-fax 4-4	Pro-fax 5-2	Pro-fax 5-4	Marlex 4050	Marlex 2350	Marlex 1 040-01
PHYSICAL										
Mold shrinkage (in./in.)--in mils		19	19	11	9	14	13			
Specific gravity		0.903	0.899	1.05	1.22	1.05	1.20	0.905	0.908	0.908
Specific volume (in.^{3}/lb)		30.68	30.81	26.38	22.7	26.38	23.08	30.61	30.51	30.51
Water absorption (%) 24 hr @ 73F (23C)		<0.02	<0.02	<0.03	<0.03	<0.03	<0.03			
equilibrium, 73F (23C)										
Haze (%)										
Transmittance (%)										
MECHANICAL										
Tensile strength (psi) yield		5,075	3,900	5,050	3,750	4,650	3,050	5,000	5,000	4,000
ultimate										
Elongation (%) yield										
rupture										
Tensile modulus (x10^5 psi)		2.5	1.85	3.77	4.22	3.37	3.24	2.25	2.4	1.6
Flexural strength (psi)										
Flexural modulus (x10^5 psi)										
Compressive strength (psi)										
Shear strength (psi)										
Izod impact strength notched,1/8 in.thick		0.8	2.5	0.6	0.6	0.9	0.8	0.7	0.5	2.2
(ft-lb/in.) unnotched,1/8 in.thick		17.5	No br.	13.5	8.9	31.6	34.0	24	15	No br.
Tensile impact strength (ft-lb/in.^{2}) S type										
L type										
Fatigue strength (psi @ ___ cycles)										
Rockwell hardness (D-Durometer)		R99	R80	R98	R85	R98	R87	D72	D73	D70
Deformation under load (%) @ 73F (23C) @ ___ F										
THERMAL										
Heat deflection temperature (F) @ 66 psi		212	177.8	257	260.6	242.6	219.2	230	235	200
@ 264 psi								135	140	130
Specific heat (Btu/lb/F or Btu/lb/C)										
Thermal conductivity (Btu/hr/ft^2/F/in.)										
Coeff. therm. exp.(x10^{-5})(in./in./F or /C)										
Brittleness temperature (F)										
Flammability UL Standard 94										
Oxygen index										
ELECTRICAL										
Dielectric strength (V/mil) 60H										
10^{6}H										
Dielectric constant 60H										
10^{6}H										
Power factor										
Volume resistivity (ohm-cm)										
Arc resistance (sec), tungsten electrodes										

Polypropylene

Item 1 is a general-purpose grade with high impact strength suitable for miscellaneous applications such as food containers by hot-fill method, brush bristles, and packaging strapping.

Item 2 is a long-term heat-resistant grade [10 years under continuous exposure at 250°F (120°C)] and retains the properties when exposed to long-term aqueous detergent/bleach solution. Applications are clothes washer and dishwasher parts.

Item 3 is a mineral-reinforced grade with high stiffness and resistance to distortion at elevated temperatures. It is frequently used for appliance and automotive components.
All the above are products of Amoco Chemicals Corp.

Item 4 is a 20% fiberglass reinforced (G.R.) and *Item 5* is a 40% G.R. grade; both are chemically coupled glass to resin. This feature improves significantly the physical properties when compared to standard glass-reinforced compounds. The improvement is attained in the area of dimensional stability and creep resistance at elevated temperatures, stiffness, impact strength, and other mechanical strengths. The utility resin is thereby raised to a level of other engineering thermoplastics. When considering materials for a product, these compounds merit full evaluation in terms of cost per cubic inch and properties and processing characteristics against other polymers.
They are products of LNP Corp.

Item 6 is a heat-stabilized grade for extra life at high temperatures coupled with toughness and favorable moldability. It is applied to closures, large parts requiring a fair amount of impact resistance, battery caps, and automotive air ducts.

Item 7 is known as a "polyallomer," i.e., a formulation based on block copolymers of propylene and ethylene. It offers these benefits over a crystalline polypropylene: a low brittleness temperature of −40°F (−40°C), impact strength as high as 12 ft-lb/in. of notch, lower notch sensitivity, improved hinging characteristics, and improved resistance to fatigue in flexing.
The usual copolymerization of propylene and ethylene causes loss of crystallinity and the result is an amorphous rubbery polymer. *Item 7* is used for shoe lasts, heavy closures, pipe fittings, and automotive items.

Item 8 has an unusual combination of stiffness and toughness. It is used for automotive parts where standard polypropylene grades are not suitable because they do not have high stiffness combined with greater toughness.

Item 9 is a talc-filled grade that has a higher modulus of elasticity, good tensile strength, better resistance to heat distortion, and lower mold shrinkage than general-purpose polypropylene.

Items 6—9 are "Tenite," a trademark of Eastman Kodak Co.

Item 10 is a general-purpose grade suitable for housewares, chemical equipment and containers, and hospital ware.

Item 11 is an improved low-temperature impact grade, and is used for totes, seating, and automotive parts.

Item 12 is an *Item 10* resin filled with 20% talc, and has maximum stiffness, outstanding heat-aging resistance, high resistance to solvents and chemicals, excellent environmental stress-cracking resistance, good dimensional stability, and low shrinkage.

Item 13 is the *Item 11* resin filled with 40% talc to provide an optimum balance of stiffness and impact resistance. Other characteristics are the same as *Item 12*.

Item 14 is an *Item 10* resin filled with 20% calcium carbonate to give a material with a high impact strength with good stiffness, good heat-aging resistance, high resistance to solvents and chemicals, excellent environmental stress-cracking resistance, and low shrinkage.

Item 15 is an *Item 11* resin filled with 40% calcium carbonate to provide the highest impact strength possible with good stiffness. Other characteristics are the same as *Item 14*.

Items 10—15 are "ProFax," a tradename of Hercules, Inc.

Item 16 has excellent long-term heat stability and resistance to wet—dry extraction. It is applicable to appliance parts, chemical equipment, and industrial parts.

Items 17 has superb processibility and good toughness and is suitable for thin-walled containers, closures, syringes, and medical supplies.

Item 18 possesses high impact strength, excellent weld line strength, and good long-term heat stability. It is used for carrying cases, chair shells, and automotive parts.

Items 16—18 are "Marlex," a trademark of Phillips Chemical Co.

Information and illustrations have been adapted from technical publications of listed companies. Additional grades are available from these and other suppliers.

Electronic applications of polypropylene.

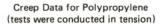

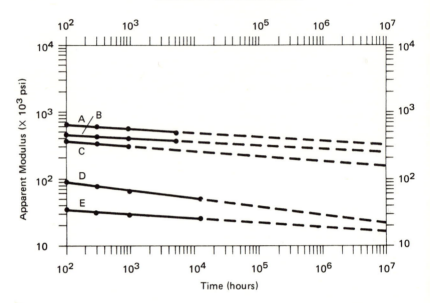

A, Profax, 30% glass reinforced at 73°F (23°C) and 4000 psi; B, Profax, 20% glass reinforced, at 73°F (23°C) and 2500 psi; C, Profax, 20% glass reinforced, at 176°F (80°C) and 1500 psi; D, Profax 6423 (unfilled) at 73°F (23°C) and 500 psi; E, Profax 6423 (unfilled) at 140°F (60°C) and 600 psi. Dotted lines indicate extrapolated values.

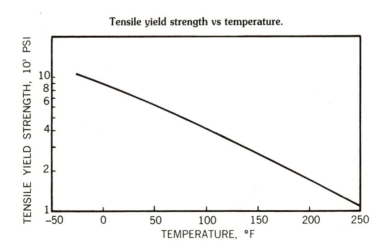

Tensile yield strength vs temperature.

Pro-fax® polypropylene shrinkage variation with thickness.

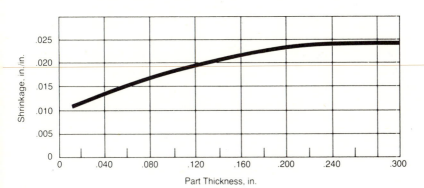

POLYSTYRENE (PS)

PS is a thermoplastic and amorphous material. It is a polymer made by the polymerization of styrene as the sole monomer. The general purpose grade is a clear material, rigid (high tensile and high modulus), creep resistant, highly notch sensitive, and resistant to water, acids, bases, alcohols, and detergents. It lacks in UV stability; therefore, it is not considered weather resistant unless suitably modified with appropriate additives. When exposed to chlorinated hydrocarbons, the surface will be marred, and, when parts are stressed, they will fail owing to this exposure. In general, exposure to aliphatic and aromatic hydrocarbons should be avoided. Even prolonged exposure to coconut oil and butter is harmful to the material.

The rubber-modified grade changes the above properties in a favorable manner. Izod impact strength is improved many fold, however, it is at the expense of tensile strength and modulus, which decrease; UV stability is also adversely affected. Tensile elongation is greater, and the rubber-modified polymer results in a tougher resin.

Structural foam material is also being produced from the rubber-modified grade. The structural foam grade has higher rigidity for equal weight, will not show sink marks, has lower internal stresses, and lends itself to manufacture of larger parts with a resulting lower cost per volume.

The rubber-modified grade is also available with a flame-retardant additive for applications such as clock and radio housings. It is to be noted that this grade is *not* suitable for contact with food.

Another specialty resin is available in the form of expandable beads containing a blowing agent. It is widely used for making low-density foamed material for packing and heat insulating purposes.

There are other special purpose grades formulated to display certain specific characteristics required in an application. Raising any one property is usually attained at the sacrifice of another. Polystyrene being a low cost material, there is continuous development work carried on to copolymerize with other monomers to achieve some desired purpose. A resin known as "K" resin, produced by Phillips Chemical Co., is a crystal clear and impact resisting material. It is a butadiene–styrene (BDS) resin, which is used wherever see-through properties combined with toughness are needed.

The applications are many, especially in the field of packaging as containers for food and produce, disposable dishes, and institutional serviceware. Drug containers, fastfood trays, egg crates, and meat and poultry containers are many of the applications for this material. The applications also extend to refrigerator liners and components, tape reels for computers, and miscellaneous electronic parts. Furniture components, housewares of various types, and novelty items also contribute to the utilization of the material.

The expandable bead material finds application in insulated disposable drinking cups, numerous cushioning shapes for shipping purposes, and a variety of containers for heat-insulation purposes.

Processing is by injection molding, blow molding, extrusion, hot forming, and other thermoplastic forming processes. The expandable bead material is manufactured in machinery designed for this grade of polystyrene.

Polystyrene

Item 1 is a general-purpose crystal grade suitable for containers, thin-section housewares, novelties, vials, disposable tumblers, and medicinal items.

Item 2 is an impact grade that has superior thermal resistance and allows the molding of thin-wall parts for higher service temperatures. The application is for video and audio cassettes, cartridges, and slide mounts.

These items are "Fosta" products, a tradename of American Hoechst Plastics.

Item 3 is a heat-resistant crystal material and *Item 4* is a high heat, high impact strength material for such applications as refrigerator door and food liners, as well as photographic cartridges and appliance components.

Items 3 and 4 are "Lustrex" products, a tradename of Monsanto Plastics and Resins Co.

Item 5 is used for containers, housewares, drinkware, and medical laboratory items.

Item 6 is applied to deep food containers, housewares, and furniture.

Items 5 and 6 are products of Amoco Chemicals Corp.

Items 7 and 8 are glass-reinforced impact polystyrene materials with 20% and 40% glass content. They are products of LNP Corp.

Item 9 is known as "K-Resin" and is unique in that it is a crystal-clear material with good impact strength, a property not found in this type of polymer. This product is tradenamed of Phillips Chemical Co.

POLYSTYRENE

PROPERTY	1. Fostarene 817	2. Fosta Tuf-Flex 1552	3. Lustrex H101	4. Lustrex 3350	5. Amoco G1	6. Amoco H4	7. LNP-CF 1004 (20% GR)	8. LNP-CF 1008 (40% GR)	9. K-resin-KR03
PHYSICAL									
Mold shrinkage (in./in.)--in mils			4-8	6			1-2	0.5	
Specific gravity	1.05	1.04	1.04	1.06	1.05	1.03	1.2	1.38	1.01
Specific volume (in.³/lb)	26.38	26.63	26.63	26.13	26.38	26.89	23.1	20.1	27.43
Water absorption (%) 24 hr @ 73F (23C)			0.03	0.07			0.07	0.05	0.09
equilibrium, 73F (23C)							0.14	0.10	
Haze (%)									
Transmittance (%)			88						90
MECHANICAL									
Tensile strength (psi) yield	5,300	6,000	7,300	3,300	6,000	2,700	11,500	15,000	3,200
ultimate						2,900			
Elongation (%) yield	1.5	25	2.2	40	1.4	50	2-3	2-3	100
rupture									
Tensile modulus (x10⁵ psi)	4.0	3.5		3.2	4.6		10.5	16.5	
Flexural strength (psi)	8,400	8,100	14,000		8,500		15,000	17,500	5,400
Flexural modulus (x10⁵ psi)	4.0	3.7			4.5	2.8	9.5	15.0	2.35
Compressive strength (psi)							16,700	17,800	
Shear strength (psi)									
Izod impact strength notched,1/8 in.thick (ft-lb/in.)	0.4	1.1	0.35	1.3		3.4	0.9	1.2	
unnotched,1/8 in.thick							2-3	2-3	
Tensile impact strength (ft-lb/in.²) S type									
L type									
Fatigue strength (psi @ ___ cycles)									
Rockwell hardness	M75	M45	M78				M90	M93	
Deformation under load (%) @ 73F (23C) @4000@122 F			0.8				0.7	0.2	
THERMAL									
Heat deflection temperature (F) @ 66 psi	198					180	220	235	
@ 264 psi	165	200	200	192		170	200	220	160
Specific heat (Btu/lb/F or Btu/lb/C)							0.28	0.24	
Thermal conductivity (Btu/hr/ft²/F/in.)							1.8	2.2	
Coeff. therm. exp.(x10⁻⁵) (in./in./F or /C)			6.8				2.2	1.6	7.8
Brittleness temperature (F)									
Flammability UL Standard 94	94HB	94HB			94HB	94HB			
Oxygen index									
ELECTRICAL									
Dielectric strength (V/mil) 60H, 10⁶H			500						
Dielectric constant 60H, 10⁶H									2.5
Power factor			0.0003						0.001
Volume resistivity (ohm-cm)			10¹⁷						

POLYURETHANE (PUR)

PUR can be formulated as a thermoplastic or a thermosetting resin. The outstanding characteristics of the resins are toughness, low-temperature flexibility, good capacity for load bearing, resistance to abrasion, and resistance to fuels, oils, and oxygen. They also have good shock absorption and good vibration dampening capabilities.

The urethanes are made from the chemical reaction of an isocyanate and a polyol. Under each one of the component headings there are a wide variety of commercial raw materials that when properly combined will react to form materials with properties to fit many applications.

There are a few classes of grades with major fields of application, but within each grade there are subdivisions with variation in density and other properties to suit individual needs.

Flexible foam is used for seat cushioning in furniture and transportation, as well as an underlay in carpeting. This foam is produced by a continuous slabstock process. For high-volume uses, molded shapes are a frequent choice.

Rigid foam is basically a construction material and is laminated to many facings for insulation of roofs and walls. It is molded onto the facing materials in a continuous laminating process that results in economical insulating boards for building industry.

Rigid foam is also used for providing the insulating shell for refrigerators and freezers. A mixed amount of material is poured between the mixer liner and outer shell by a mixing device, and this assembly remains in position until cured and removed to allow processing of the following unit. The excellent heat-insulating properties of the rigid foam have extended its application to water heaters, refrigerated trailers, tank cars, and boxcars.

Semirigid foam is formulated to produce parts either with an integral skin or molded against a formed sheet of ABS or vinyl to form the skin or molded without the skin effect for shock absorption; finally, it is used to produce shoe soles. The shapes can be molded in open forms or closed molds. The applications are automotive instrument panels that provide impact resistance and a soft feel. Bicycle seats and automotive steering wheels use the integral skin formulation. The shock-absorbing grade is used for bumper parts that will absorb crash energy.

The shoe sole application is dictated by the lightweight abrasion resistance, flexibility at normal and cold temperatures, and general comfort provided.

Reaction-injection-molded (RIM) parts are produced with an integral skin in a one-step operation that begins with basic raw material and ends with the finished molded part within a 2-minute cycle. The process was originally developed to make large urethane front and rear parts for automobiles to absorb crash energy. The process uses the basic component chemicals that are precisely metered in volume, mixed at high velocities, and injected into the mold where the mixture expands and fills out the mold configuration and cures to produce a finished product. The potential of the process is very favorable not only for urethane but also for other materials simply because it eliminates "in-between" steps, thus making it a most economical molding method. The proper choice of ingredients can produce parts with properties that meet automotive needs.

This same process can be used to produce urethane structural foam parts to increase rigidity in the same volume of material.

Thermoplastic Urethane (TPU)

TPU is of greatest interest to the designer in comparison with the grades previously described. TPU is formed by the reaction of bifunctional polyols with diisocyanates. There is a cross-linking effect in the basically linear material, but this cross-linking is reversible, and any material waste associated with the production of parts, but not included in the finished product, can be reprocessed.

The characteristics ascribed to the family of polyurethanes also apply to TPU. In this grade we are also dealing with many raw material ingredients that result in materials with preferred properties for specific uses. The ultimate mechanical properties are attained through a combination of molding (extruding) and room-temperature conditioning.

The applications include wheels for heavy machinery, gears, and drive belts. In the automotive industry, fender extenders, sight shields, shock-absorber parts, gas line tubing, and miscellaneous mechanical parts are made from TPU. Grades are also available for flammability and UV resistance, as well as those suitable for contact with food and medicine.

Casting grades that form thermosetting products are available in a variety of formulations that will result in parts with needed properties.

Some of the applications are chutes, gaskets, industrial truck tires, seals, and conveyor belts.

Polyurethane

The various grades of urethane are formulated to provide abrasion resistance, favorable coefficient of friction, resilience, tear resistance, a range of shore hardnesses, a specified "set" after a designated elongation or compression, resistance to permeability, cushioning, etc. If we recognize that each one of these properties has numerous subdivisions and they, in turn, by themselves or in combination with any one of a number of others will result in a multitude of grades, we can readily see that this material category does not lend itself to the treatment outlined in the data sheets.

Generally speaking, the urethane materials can be better evaluated using the same guidelines as for rubber.

POLYVINYLCHLORIDE (PVC)

PVC is a thermoplastic material. It is prepared by the polymerization of vinyl chloride as the sole monomer. PVC homopolymers and copolymers can be compounded to produce a broad range of physical properties. During compounding into a commercial resin, ingredients are added, such as stabilizers to prevent heat degradation, plasticizers to impart flexibility, lubricants as processing aids, fillers for cost-reduction purposes, impact modifiers to improve toughness, and processing aids to improve the processing. Each of the additives represents a variety of materials and those, in turn, in various combinations with each of the other additives can result in large number of grades of different properties.

The range of applications covers rigid pipe and building materials on one end and thin and flexible surgeons' inspection gloves on the opposite end.

Rigid PVC has good dimensional stability, chemical inertness, high impact resistance, higher-temperature resistance, good corrosion resistance, thermal and electrical insulation, weather resistance, clarity, and colorability.

Its applications are in water supply and sanitary systems, duct systems, house siding, window sash, gutters, flashing, and electrical conduits. The flexible and semiflexible grades are applied to all sorts of products such as furniture, briefcases, and floor tile.

Processing is done by injection molding, by extrusion into sheets, film, and shapes, by blow-molding, by thermoforming, and by the other processes employed in thermoplastics.

Polyvinylchloride

"Geon" is a tradename of B. F. Goodrich Co.

Item 1 is a general-purpose grade with average impact strength.

Item 2 is a high impact strength material.

Item 3 is transparent.

Item 4 is for exterior applications.

The data are adapted from commercial publications.

POLYVINYLCHLORIDE (PVC)

	PROPERTY	GRADE	1. Geon 827237	2. Geon 87238	3. Geon 87242	4. Geon 85856
PHYSICAL	Mold shrinkage (in./in.)--in mils		4-6	4-6	4-6	3-5
	Specific gravity		1.4	1.35	1.35	1.45
	Specific volume (in.^{3}/lb)		19.79	20.52	20.52	19.10
	Water absorption (%) 24 hr @ 73F (23C)					
	equilibrium, 73F (23C)					
	Haze (%)					
	Transmittance (%)					
MECHANICAL	Tensile strength (psi) yield		6,000	5,900	7,650	6,400
	ultimate					
	Elongation (%) yield					
	rupture					
	Tensile modulus (x10^5 psi)		3.67	3.25	4.15	4.6
	Flexural strength (psi)		11,500	11,000	13,400	10,000
	Flexural modulus (x10^5 psi)		4.05	3.72	4.65	4.35
	Compressive strength (psi)					
	Shear strength (psi)					
	Izod impact strength notched, 1/8 in.thick		1.10	15.0	2.3	13.0
	(ft-lb/in.) unnotched, 1/8 in.thick					
	Tensile impact strength (ft-lb/in.^{2}) S type					
	L type					
	Fatigue strength (psi @ ___ cycles)					
	Rockwell hardness					
	Deformation under load (%) @ ___ @ 73F (23C)					
	@ ___ @ ___ F					
THERMAL	Heat deflection temperature (F) @ 66 psi					
	@ 264 psi		154	153	152	160
	Specific heat (Btu/lb/F or Btu/lb/C)					
	Thermal conductivity (Btu/hr/ft^2/F/in.)					
	Coeff. therm. exp.(x10^{-5}) (in./in./F or /C)					
	Brittleness temperature (F)					
	Flammability UL Standard 94					
	Oxygen index					
ELECTRICAL	Dielectric strength (V/mil)					
	Dielectric constant 60H					
	10^{6}H					
	Power factor 60H					
	10^{6}H					
	Volume resistivity (ohm-cm)					
	Arc resistance (sec), tungsten electrodes					

SILICONE

The silicones are unusual polymers in that they are partly organic and partly inorganic. They are based on polymers in which the main polymer chain consists of alternating silicone and oxygen atoms. They are derived from sand and methyl chloride.

The characteristics that are responsible for most of the applications are good release properties, high degree of lubricity, excellent water resistance, good electrical properties, chemical inertness, retention of most properties at elevated temperatures compatible with the human body, and weather resistance. These properties can be varied in degree, and the physical form in which they are available is determined by molecular weight, extent of cross-linking between molecular chains, and type as well as number of organic groups attached to the silicone atoms.

Molding compounds are available in which the silicone resin is a binder for a variety of fillers that can be molded into rigid parts. The application is for encapsulation of semiconductor devices where favorable electrical and thermal properties as well as protection against moisture and good flammability resistance are important requirements.

A combination of silicone and epoxy compounds is also available for cases where the silicone characteristics combined with the strength of epoxy are needed. Silicones are utilized as the binder for laminated materials such as glass cloth, mica, and asbestos cloth for a variety of electronic board applications.

A flexible grade of silicone is available for pouring a gasket in place that will seal and cement parts together while cure takes place at room temperature (RTV—room temperature vulcanizing). The joint created by such a gasket remains flexible over a prolonged period of time and under most adverse environments.

Another flexible grade is used for casting flexible molds over models. These molds are suitable for sample parts. This same material is also used in industry for making flexible molds and patterns.

Silicone rubber parts can be made by compression, injection, and transfer molding and are used in industry where the superior qualities of the silicones are needed. Examples are high-voltage lead wire insulators, seals, connector plugs, "O" rings, and diaphragms.

The use of the material for part release, as an aid to flow, and for inherent release of plastic material is quite extensive.

SILICONE LAMINATES PROPERTY	GRADE	1. Glass fabric base	2. Asbestos fabric base
PHYSICAL			
Mold shrinkage (in./in.)--in mils			
Specific gravity		1.75	1.75
Specific volume (in.^{3}/lb)		15.9	15.9
Water absorption (%) 24 hr @ 73F (23C)		0.32	1.25
equilibrium, 73F (23C)			
Haze (%)			
Transmittance (%)			
MECHANICAL			
Tensile strength (psi) yield			
ultimate		23,000	
Elongation (%) yield			
rupture			
Tensile modulus (x10^5 psi)		17.5	
Flexural strength (psi)		24,000	14,000
Flexural modulus (x10^5 psi)		22.0	
Compressive strength (psi)		35,000	45,000
Shear strength (psi)		18,000	
Izod impact strength notched, 1/8 in.thick		9.0	7.5
(ft-lb/in.) unnotched, 1/8 in.thick			
Tensile impact strength (ft-lb/in.^{2}) S type			
L type			
Fatigue strength (psi @ ___ cycles)			
Rockwell hardness		M100	
Deformation under load (%) @ ___ @ 73F (23C)			
@ ___ @ ___ F			
THERMAL			
Heat deflection temperature (F) @ 66 psi			
@ 264 psi			
Specific heat (Btu/lb/F or Btu/lb/C)		0.26	
Thermal conductivity (Btu/hr/ft^2/F/in.)		7.0	
Coeff. therm. exp.(x10^{-5}) (in./in./F or /C)			
Brittleness temperature (F)			
Flammability UL Standard 94			
Oxygen index			
ELECTRICAL			
Dielectric strength (V/mil)		330	
Dielectric constant 60H		4.8	
10^{6}H		4.0	
Power factor 60H		0.01	
10^{6}H		0.005	
Volume resistivity (ohm-cm)			
Arc resistance (sec), tungsten electrodes		150	

STYRENE–ACRYLONITRILE (SAN)

SAN is a thermoplastic material and a copolymer of styrene and acrylonitrile.

The main advantage it possesses over polystyrenes is chemical resistance, toughness, load-bearing property, and somewhat higher heat deflection temperature. SAN additionally has a high modulus and good creep resistance.

This material resists nonoxidizing acids, aliphatic hydrocarbons, alkalis, battery acids, vegetable oils, foods, and many detergents. It is attacked by aromatic and chlorinated hydrocarbons, esters, and ketones.

Applications cover component parts for refrigerators (meat and vegetable compartments); parts for blenders and mixers, such as containers for ingredients; battery cases; tumblers; and several automotive components.

Processing includes injection molding, extrusion, and blow molding.

Styrene–Acrylonitrile

"Lustran" is a tradename of Monsanto Plastics and Resins Co.

Item 1 is a general purpose grade and *Item 2* a UV stabilized grade.

Items 3–5 are glass-reinforced SAN by LNP Corp. and are being applied for automotive instrument panels, crash pad retainers, and business machine and photographic equipment components.

Additional grades are available from the above and other suppliers. The data have been adapted from commercial publications.

STYRENE-ACRYLONITRILE (SAN) PROPERTY		1. Lustran SAN 31	2. Lustran SAN 31-7000	3. LNP-BF1004 (20% GR)	4. LNP-BF1006 (30% GR)	5. LNP-BF1006 (40% GR)
PHYSICAL	Mold shrinkage (in./in.)--in mils	3-4	3-4	1-2	0.5	0.5
	Specific gravity	1.07	1.07	1.22	1.31	1.40
	Specific volume (in.^{3}/lb)	25.89	25.89	22.7	21.2	19.8
	Water absorption (%) 24 hr @ 73F (23C)	0.25	0.25	0.15	0.1	0.08
	equilibrium, 73F (23C)			0.36	0.32	0.28
	Haze (%)	1.4	1.5-1.8			
	Transmittance (%)					
MECHANICAL	Tensile strength (psi) yield					
	ultimate	10,500	10,500	15,800	17,400	18,600
	Elongation (%) yield					
	rupture	3.0	3.0	2-3	2-3	2-3
	Tensile modulus (x10^5 psi)	4.75	4.75	11.5	16.0	19.5
	Flexural strength (psi)	11,500	11,000	19,800	22,000	23,200
	Flexural modulus (x10^5 psi)	5.0	5.0	11.0	15.0	18.5
	Compressive strength (psi)			19,400	21,000	23,200
	Shear strength (psi)			9,000	9,400	9,800
	Izod impact strength notched, 1/8 in.thick	0.45	0.45	1.0	1.0	1.1
	(ft-lb/in.) unnotched, 1/8 in.thick			3-4	3-4	3-4
	Tensile impact strength (ft-lb/in.^{2}) S type					
	L type					
	Fatigue strength (psi @ ___ cycles)					
	Rockwell hardness	M83	M83	R122	R123	R123
	Deformation under load (%) @ ___ @ 73F (23C)					
	@4000 @ 122 F	1.5	1.5			
THERMAL	Heat deflection temperature (F) @ 66 psi			220	230	235
	@ 264 psi	205	205	205	215	220
	Specific heat (Btu/lb/F or Btu/lb/C)			0.21	0.25	0.28
	Thermal conductivity (Btu/hr/ft^2/F/in.)			1.8	2.0	2.2
	Coeff. therm. exp.(x10^{-5}) (in./in./F or /C)			2.1	1.8	1.5
	Brittleness temperature (F)					
	Flammability UL Standard 94			94HB	94HB	94HB
	Oxygen index					
ELECTRICAL	Dielectric strength (V/mil)					
	Dielectric constant 60H					
	10^{6}H					
	Power factor 60H					
	10^{6}H					
	Volume resistivity (ohm-cm)					
	Arc resistance (sec), tungsten electrodes					

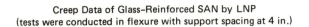

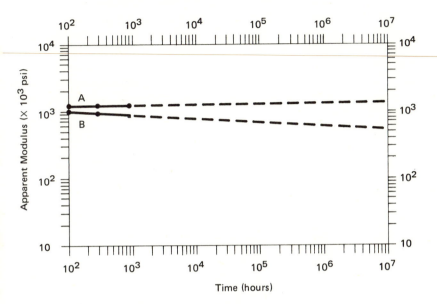

A, BF 1006, 30% glass–reinforced SAN at 75°F (24°C) and 5000 psi; B, BF 1004, 20% glass–reinforced SAN at 75°F (24°C) and 5000 psi. Dotted lines indicate extrapolated values.

SULFONE POLYMERS

The sulfone materials are thermoplastic and amorphous. There are three polymers in the sulfone family of resins, which will be described in the following paragraphs. This group of polymers has such common characteristics as transparency, high heat resistance, dimensional stability, rigidity, high strength, toughness, good creep resistance, excellent electrical properties over a broad range of temperatures, low moisture absorption, good flammability ratings, and hydrolysis resistance. They will also withstand prolonged exposure to higher temperature without ill effects on properties.

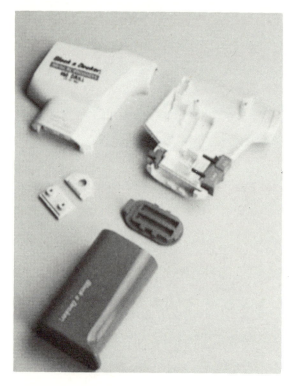

Udel polysulfone components of a surgical drill made by Black & Decker.

Sulfone Polymers

"Radel" and "Udel" are tradenames for polyphenyl sulfone and polysulfone, respectively, of Union Carbide Corp. All filled grades are products of LNP Corp. and are indicated by their numbering system.

The data and illustrations are from commercial publications of the above companies and are adapted from them.

SULFONE POLYMERS

PROPERTY	LNP-GF1006 (% GR)	LNP-GC1006 (% carbon, per filled)	Udel P1700	PPS-OF1006 (% GR)	PPS-OC1006 (% carbon, per filled)	PPS-Radel filled	PES-JF1008 (% GR)	PES-JF1004 (% GR)	PES unfilled
PHYSICAL									
Mold shrinkage (in./in.)--in mils	2-3	1-2	7	2	1	8	1.5	3	7
Specific gravity	1.45	1.37	1.24	1.56	1.45	1.34	1.68	1.51	1.37
Specific volume (in.^{3}/lb)	19.1	20.22	22.34	17.76	19.1	20.65	16.61	18.38	20.25
Water absorption (%) 24 hr @ 73F (23C)	0.20	0.15	0.3	0.04	0.04	0.05	0.31	0.37	0.43
Water absorption equilibrium, 73F (23C)									
Haze (%)									
Transmittance (%)									
MECHANICAL									
Tensile strength (psi) yield			10,200			10,400			12,200
Tensile strength (psi) ultimate	18,000	23,000		20,000	27,000		22,000	18,000	
Elongation (%) yield			50-100			7			
Elongation (%) rupture	3-4	2-3		3-4	2-3	60	3-4	3-4	
Tensile modulus (x10^5 psi)			3.6			3.1			
Flexural strength (psi)	24,000	32,000	15,400	29,000	34,000	12,400	30,000	25,000	18,700
Flexural modulus (x10^5 psi)	12.0	20.5	3.9	16.0	24.5	3.3	16.0	8.5	3.7
Compressive strength (psi)			10,200						
Shear strength (psi)	9,500	9,500							
Izod impact strength notched, 1/8 in. thick (ft-lb/in.)	1.8	1.2	1.3	1.4	1.1	1.2	1.6	1.4	1.6
Izod impact strength unnotched, 1/8 in. thick (ft-lb/in.)	14	6-7		8-9	5-6		12.0	8.0	
Tensile impact strength (ft-lb/in.^{2}) S type			200				60	40	
Tensile impact strength L type									
Fatigue strength (psi @ cycles)									
Rockwell hardness			M69				M98	M98	M88
Deformation under load (%) @ 73F (23C) @ 66 psi									
Deformation under load @ 264 psi									
THERMAL									
Heat deflection temperature (F)	365	365	345	500	505	400	420	410	395
Specific heat (Btu/lb/F or Btu/lb/C)						0.28			
Thermal conductivity (Btu/hr/ft^2/F/in.)	2.2	5.5	1.8	2.8	5.2		2.6	2.3	1.9
Coeff. therm. exp.(x10^{-5})(in./in./F or /C)	1.4	0.6	3.1	1.3	0.6	3.1	1.6	2.0	3.1
Brittleness temperature (F)									
ELECTRICAL									
Flammability UL Standard 94	94V-0	94V-0	94V-0	94V-0	94V-0	94V-0	94V-0	94V-0	94V-0
Oxygen index			30%			38%			
Dielectric strength (V/mil) 60H	480		425			371			
Dielectric strength (V/mil) 10^{6}H									
Dielectric constant 60H	3.55		3.07			3.44			
Dielectric constant 10^{6}H	3.49					3.45			
Power factor 60H	0.0019		0.0008			0.00058			
Power factor 10^{6}H	0.0049		0.0034			0.0076			
Volume resistivity (ohm-cm)	10^{17}		5x10^{16}			3.5x10^{15}			
Arc resistance (sec), tungsten electrodes	115		122						

Polysulfone tensile stress–strain curves.

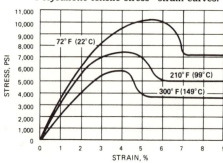

Tensile creep modulus of polysulfone vs time

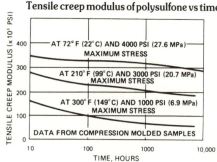

Flexural modulus vs temperature.

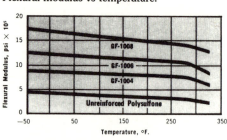

Tensile creep at 210°F (99°C).

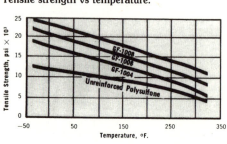

Tensile strength vs temperature.

Tensile creep modulus vs time at 210°F (99°C).

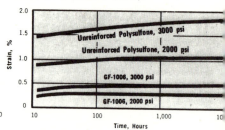

Maximum working stresses recommended for polysulfone.

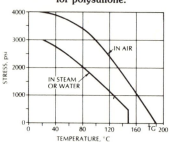

Polyethersulfone (PES)

The continuous service temperature of PES is 400°F (204°C). It is important to note that mechanical and electrical properties are practically similar to those of room-temperature results. Creep at this high temperature is also low. These favorable performances are predicated on parts being stress-relieved.

When subjected to outdoor use, carbon black should be used to improve UV resistance. The polymer is notch-sensitive, but attention during design can overcome this drawback. It is resistant to many chlorinated and fluorinated cleaning solvents and most oils and fuels. It is attacked by ketones, esters, and some halogenated and aromatic hydrocarbons. It is resistant to sterilization effects (steam, chemical, and radiation); therefore, it is used for medical accessories and equipment components.

In the electrical field PES is used for coil forms, terminal blocks, and switches when high temperatures are encountered. In aircraft applications PES is used for interior parts, radomes, and indicator lights. PES can be injection molded, extruded, and thermoformed.

Polyphenyl Sulfone

Polyphenyl sulfone is a very high impact strength material with Izod value of 12 ft-lb/in. Environment stress cracking resistance is outstanding and superior to the other sulfone grades. Its creep resistance is also superior.

Applications are for high-temperature coil bobbins, aircraft window reveals, and electrical components.

Processing is the same as for PES.

Polysulfone

Polysulfone is approved for continuous use at 300°F (149°C) and will withstand even higher temperatures for intermittent service. When exposed to 300°F (149°C) temperature for longer periods of time, the material is annealed and looses some properties early during the exposure, but following this, all the properties remain essentially constant.

Its high resistance to creep allows loading up to 4000 psi without excessive elongation. Polysulfone is notch-sensitive and has a low shrinkage and a low coefficient of expansion. Electrical properties are outstanding and are maintained at high temperatures; even after exposure to high humidity or immersion in water, the electrical properties are retained up to 350°F (177°C).

Polysulfone can be plated by an electroless process with copper and nickel. Weather resistance can be improved with incorporation of carbon black.

Polysulfone is attacked by ketones, esters, and chlorinated and armomatic hydrocarbons; it resists acids, alkali, salt solutions, detergents, oils, and alcohols.

Polysulfone's heat resistance, abrasion resistance, toughness, and high strength make it suitable for applications in processing equipment, for ovenware, and for dishware for consumer food handling. Medical instrumentation is another use for the material in view of its resistance to sterilization environments.

The outstanding dimensional stability recommends the material for wristwatch cases and automotive and space components.

Processing is done by injection, extrusion, and thermoforming.

INDUSTRIAL LAMINATES

Industrial laminates are a group of materials that can be of interest to product designers. These materials' properties have been well established over a period of many years, and they are mostly manufactured to NEMA (National Electrical Manufacturers Association) specifications.

Essentially they consist of sheet-material layers impregnated with a suitable resin and compression molded into boards of insulating materials. The size of the boards is determined by the press size and the number of sheet layers, which, in turn, are governed by use requirements. The resins, which may incorporate additives for arc resistance, lubricity, flame retardance, etc., are phenolics, melamine–formaldehyde, epoxies, silicones, polyimide binders, and polytetrafluorethylenes. The most common sheet materials used are cellulose or asbestos paper, cotton cloth or mat, linen cloth, asbestos cloth or mat, nylon cloth, and glass cloth or mat.

The materials have been applied extensively in power-generating and distribution systems for supporting and insulating switching systems, all types of meters, and other electrical components. Other applications are in steel mills and miscellaneous machine tools for drive gears to provide shock absorption and noise reduction, and also in bearings (with lubricity additives) for self-lubricating and noise reduction purposes.

Thin laminates are used for decorative and functional purposes such as sink and table tops and other structural pieces. These decorative laminates are usually cemented to wood backings.

Printed circuit boards in which copper foil is one of the sheet layers utilize laminated materials because of their high strength and favorable electrical properties.

Vulcanized Fiber

Vulcanized fiber was one of the first laminates. It consists of chemically solvated paper layers fused together to form a tough, resilient, lightweight material that has high impact strength and that will resist abrasion, scuffing, and denting. It can be punched, machined, and shaped.

Vulcanized fiber is used for tote pans, luggage, military trunks, washers, gaskets, and insulation in motors and transformers. Its arc resistance properties are utilized in fuse tubes and lightning arresters.

Industrial Laminates

The data on laminates represent average values. When considering a laminate for an application, the information should be identified for a grade on hand. The variables that contribute to a change in properties are the large number of fabrics that may be used, along with variation in resins that are present as binders.

The laminates are sold under such tradenames as Textolite, Micarta, and Formica.

INDUSTRIAL LAMINATES

	GRADE	1. Melamine-Cellulose Paper	2. Melamine-Glass Fabric	3. Phenolic-Wood Base	4. Phenolic-Cellulose Paper	5. Phenolic-Cotton Fabric	6. Thermosetting Polyester-Glass-Matt	7. Vulcanized Fiber
PHYSICAL	Mold shrinkage (in./in.)--in mils							
	Specific gravity	1.48	1.90	1.3	1.34	1.35	1.75	1.18
	Specific volume (in.^{3}/lb)	18.5	14.58	21.3	20.7	20.6	16.5	23.47
	Water absorption (%) 24 hr @ 73F (23C)	1.5	1.3	6.5	3.1	2.55	1.15	45.5
	Water absorption (%) equilibrium, 73F (23C)							
	Haze (%)							
	Transmittance (%)							
MECHANICAL	Tensile strength (psi) yield							
	Tensile strength (psi) ultimate	17,500	44,000	24,000	14,000	11,500	22,000	10,000
	Elongation (%) yield							
	Elongation (%) rupture							
	Tensile modulus ($\times 10^5$ psi)		22.5	37	14	10	20	10
	Flexural strength (psi)	17,000	60,000	32,000	20,000	22,000	22,500	16,500
	Flexural modulus ($\times 10^5$ psi)		28.0		11.5	9.0	17.5	8.5
	Compressive strength (psi)	39,000	55,000	16,500	30,000	37,000	32,500	30,000
	Shear strength (psi)		27,000		10,000	12,500	15,000	
	Izod impact strength notched, 1/8 in.thick (ft-lb/in.)	0.9	10.0	6.0	0.68	1.9	16.25	2.0
	Izod impact strength unnotched, 1/8 in.thick (ft-lb/in.)							
	Tensile impact strength (ft-lb/in.^{2}) S type							
	Tensile impact strength L type							
	Fatigue strength (psi @ ____ cycles)							
	Rockwell hardness	M118	M120	M98	M95	M98	M90	
	Deformation under load (%) @ 73F (23C) @ 66 psi							
	Deformation under load (%) @ ____ F @ 264 psi							
THERMAL	Heat deflection temperature (F)							
	Specific heat (Btu/lb/F or Btu/lb/C)		0.3	0.3	0.4	0.35	0.25	0.4
	Thermal conductivity (Btu/hr/ft^2/F/in.)		9.5		4.4	4.5	5.0	4.0
	Coeff. therm. exp.($\times 10^{-5}$)(in./in./F or /C)							
	Brittleness temperature (F)							
	Flammability UL Standard 94							
	Flammability Oxygen index							
ELECTRICAL	Dielectric strength (V/mil) 60H	550	400	275	650	375	500	225
	Dielectric strength (V/mil) 10^6H							
	Dielectric constant 60H	7.0	8.0	5.0	6.0	7.5	5.2	
	Dielectric constant 10^6H				5.1	6.25	4.5	
	Power factor 60H	0.4	0.18	0.11	0.06	0.045	0.038	
	Power factor 10^6H				0.05	0.075	0.024	
	Volume resistivity (ohm-cm)							

THERMOPLASTIC ELASTOMERS (TPE)

Thermoplastic elastomers can be defined as materials that at room temperature and under low stress will elongate to at least twice their original length, but upon the release of the stress will snap back in rubberlike fashion to their original dimensions.

These materials have physical properties similar to those of rubbers, but have the advantage of being able to be processed on conventional injection machines at cycles that are a fraction of the vulcanized rubber curing time. In addition to this, any scrap generated during the molding operation can be remolded. In these days of high material costs, this is another important favorable factor. The factors in comparison with rubber are lower elongation and higher permanent set when compressed for prolonged times such as is the case with gaskets, "O" rings, and other sealing agents.

There are three families of TPE and they are

1. Styrene/elastomer block copolymer

2. Urethane block copolymer and polyester/polyether block copolymers

3. Polyolefin blends

The term *block copolymer* covers essentially a linear copolymer in which there are repeated sequences of polymeric parts of a different chemical composition.

The thermoplastic part in the copolymer has a cross-linking effect on the rubbery part of the compound and acts like a chemically cross-linked system. In the TPE compound, however, the structure is reversible in contrast to a chemical cross-linked composition.

As a group, the TPEs have certain common characteristics. They are flexible even at low temperatures, tough, resist the effects of weather as well as a variety of solvents and chemicals, and paintable.

The applications of the different grades of TPE will provide an indication as to their specific properties.

Grade #1 is used for tubing, hose, insulation for wire and cable, and shoe soles. Flexibility, resistance to abrasion and flex cracking are the features that make this grade attractive for use as shoe soles.

Grade #2 is used extensively in the automotive industry for bumper fillers, fender extensions, sight shields, grille components, and decorative parts. Hydraulic hose, gears, semipneumatic tires, and flexible coupling parts also use this grade.

Grade #3 is also applied to automotive parts as in Grade #2 uses. Most of this grade is utilized as electrical insulation for extruded wire and cable.

Processing can be carried out by all methods employed with thermoplastics.

Thermoplastic Elastomers

Thermoplastic elastomers are, as a rule, substitutions for rubber products. If they are to be successful as a replacement, they should be evaluated along the same lines as the materials that they replace. Some of the guidelines are: (1) resistance to permeability to water and gases, (2) ability to withstand large deformation without serious structural damage, (3) ability to absorb the energy of shock by mechanical hysteresis, (4) resistance to tear, and (5) "permanent set" under varying conditions of usage.

The data sheet developed for this material does not cover any of the above requirements; therefore, it is omitted.

PRELIMINARY MATERIAL CONSIDERATIONS FOR PRODUCTS

After the commercially available polymers have been reviewed with respect to property characteristics and data, the next item of interest is to make a comparative evaluation of thermoset and thermoplastic materials. The comparison will be made along the lines of permanence-related performance, heat-related properties, and overall cost effectiveness.

This evaluation should alert the designer to the fact that for some properties one type is preferred over the other, and therefore deserves a complete analysis before considering another type.

Thermosetting Type	Thermoplastic Type
Permanence Performance Characteristics	
Thermosets undergo a chemical change during molding, becoming solid, and therefore, they cannot be melted.	Thermoplastics are softened by heat, and become solid on cooling. This process can be repeated many times.
Overall chemical resistance of thermosets is unequalled by any thermoplastic.	Chemical resistance of thermoplastics is limited to fewer compounds than is the case with thermosets.
Dimensional stability is excellent.	Since thermoplastics respond more readily to the influence of heat, they also tend to fluctuate in dimensions. Some polymers vary more than others, but, as a group, they would rank second to thermosets.
Creep over a prolonged time period is well below even glass-reinforced engineering thermoplastics. This means better dimensional stability and better resistance to property deterioration.	Creep is one of the major problems with the thermoplastics when subjected to long-term loads. This means sizable dimensional changes and strength degradation. Creep resistance is improved by fillers and reinforcements.
Molded-in stresses with compression-molded parts are lower than in other molded parts, and where minimal distortion is a factor, it can be a deciding consideration.	Injection-molded thermoplastics have molded-in stresses in varying degrees that are caused by part design, processing parameters, and mold design.
Toughness is a property that can be found in thermosets at a considerable	Most thermoplastics are inherently tough materials and for this reason are

Thermosetting Type	Thermoplastic Type

Permanence Performance Characteristics (*Continued*)

cost by using reinforcing materials such as fabrics and/or fiberglass mat.

used whenever this requirement is needed. They provide good toughness at low cost.

Colors in thermosetting compounds are limited in variety and their stability is not satisfactory. The resins tend to discolor over prolonged periods of time.

Thermoplastic materials can be colored to any desired shade of appearance and normally maintain the color throughout the life of a product.

Very few thermosets are available in clear, see-through materials.

The selection of clear material in thermoplastics is quite large.

Shrinkage of compression-molded parts is low, but intermediate for transfer and injection-molded pieces.

Shrinkage for crystalline and semicrystalline materials is high, but for amorphous compounds, it is low.

Thermosets have been used a great many years and under varying conditions, so that anticipated required functions can be accurately predicted.

The application of thermoplastics to engineering products is relatively new. Their behavior under load over prolonged periods of time frequently has to be interpolated on the basis of short-term test data. The design of thermoplastic parts should therefore be carried out with greater caution.

Heat-Related Characteristics

Heat deflection temperature is considerably higher than most so-called engineering thermoplastics.

Heat deflection temperature varies with each generic thermoplastic and indicates the start of material softening.

Use temperature limit can be almost double that of most thermoplastics which enjoy wide applications.

Use temperature, because of a tendency to soften under heat, is relatively low.

Rigidity in relation to elevated temperature is constant up to at least 250°F (121°C).

Theoretically, as well as practically, rigidity starts declining as soon as the temperature increases above 73°F (23°C).

Coefficient of thermal expansion is about one-half that of thermoplastics.

Many thermoplastics are used in conjunction with other materials, and the high coefficient of thermal expansion can present use problems.

Thermosetting Type	Thermoplastic Type

Heat-Related Characteristics (*Continued*)

Thermosetting Type	Thermoplastic Type
Flammability resistance is inherent to the thermosets.	Thermoplastics enhance their flammability resistance by additives.

Cost Effectiveness Characteristics

Thermosetting Type	Thermoplastic Type
Processing costs are or have been higher than thermoplastics. With improved materials and advanced technology, the manufacturing price is approaching that of the thermoplastics.	Processing of thermoplastics is easier and requires less skill. In most cases, they will have an edge over thermosets when comparing manufacturing.
Mold wear is greater with thermosets. They not only require better tool steels and heat treatment, as well as chrome plating to resist wear, but are also prone to damage where flash is accidentally left at the parting line.	Unfilled and unreinforced thermoplastic molds will produce many times greater quantities of parts even when the tool is made of semihard steel (prehardened). There is no tendency to generate flash, thus the danger of mold damage is minimized.
Material waste connected with processing has been, in the past, a total loss. At present, methods of reuse of waste as a filler are being developed. This should somewhat decrease the waste factor in the material.	Thermoplastic loss of material during molding operations is reusable. With the continuous increase in resin price, this is an important advantage.
Specific gravity is usually higher for thermosets and occasionally is detrimental to their application.	Most of the thermoplastics have a lower specific gravity, and where weight is a factor, this can be a valuable advantage.
Cost per cubic inch should, in many cases, be lower, since the most expensive ingredient makes up about one-half of the compound, while the remainder is a low-cost filler.	Thermoplastics, although lighter in weight, should be more expensive per cubic inch since they are mostly used in the form of 100% resin.
Production cycles with screw-preheating systems (injection or compression) compare favorably with thermoplastics. On the other older methods they can be 50%–100% longer than thermoplastics.	Effective mold temperature control can contribute to very short cycles of production with thermoplastics, which overall is faster than thermosetting molding.

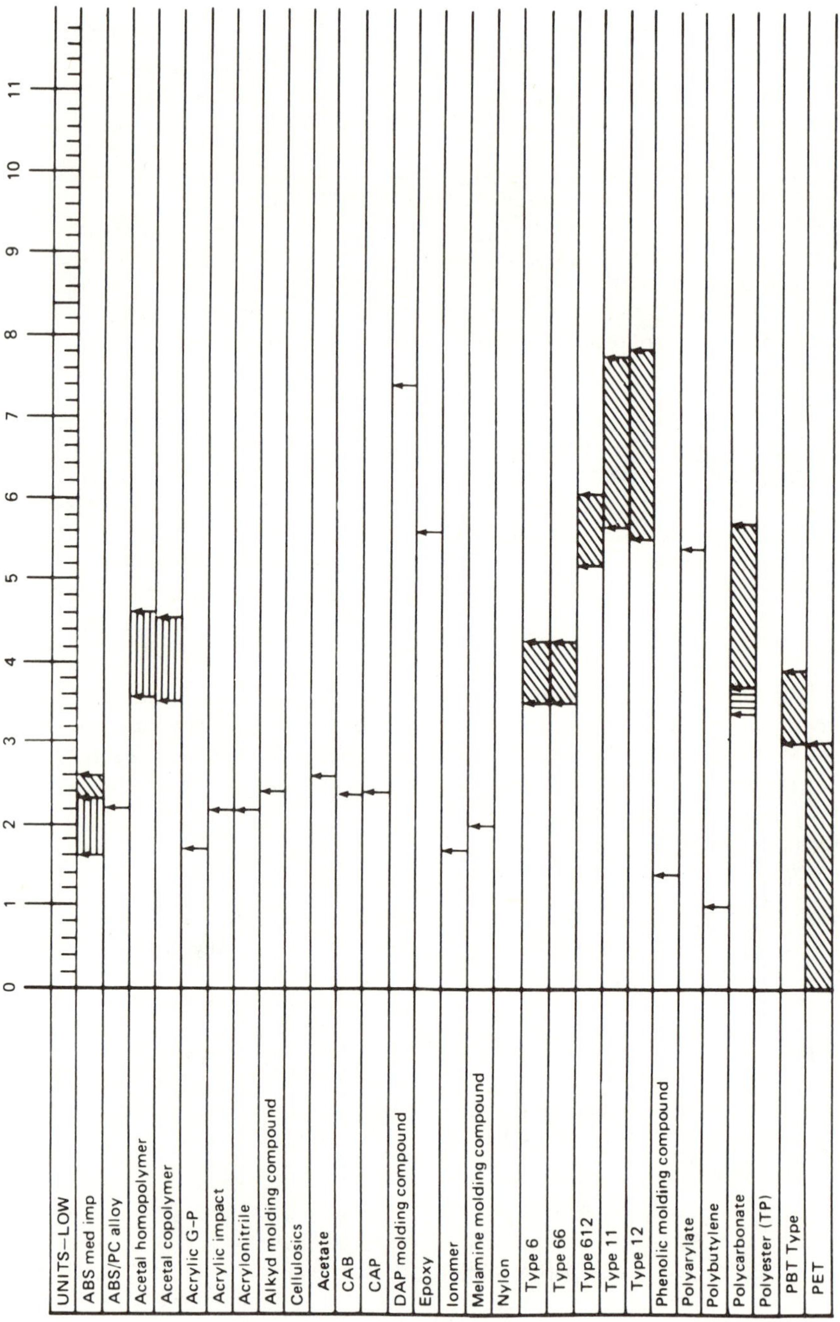
UNITS—LOW
ABS med imp
ABS/PC alloy
Acetal homopolymer
Acetal copolymer
Acrylic G-P
Acrylic impact
Acrylonitrile
Alkyd molding compound
Cellulosics
Acetate
CAB
CAP
DAP molding compound
Epoxy
Ionomer
Melamine molding compound
Nylon
Type 6
Type 66
Type 612
Type 11
Type 12
Phenolic molding compound
Polyarylate
Polybutylene
Polycarbonate
Polyester (TP)
PBT Type
PET
0
1
2
3
4
5
6
7
8
9
10
11

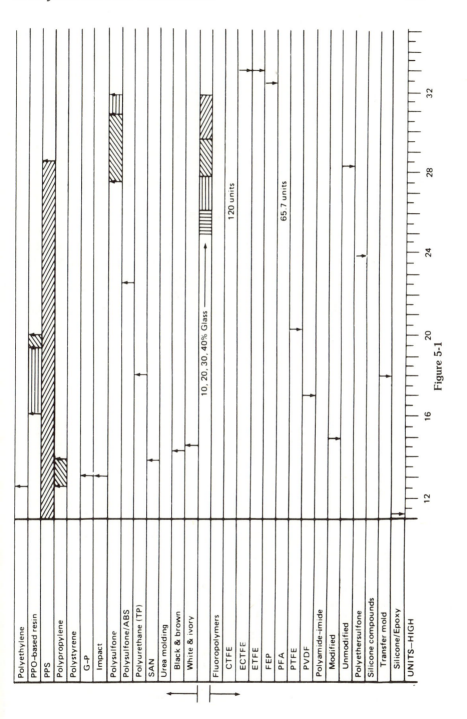

Figure 5-1

Thermosetting Type	Thermoplastic Type
Cost Effectiveness Characteristics (*Continued*)	
Compression-molded parts usually require deflashing. This presents a slight added cost.	No flash is produced; therefore, no cost is involved.

The value analysis of thermoplastic and thermoset materials provides one guidepost for material selection. The second criterion that should be considered is the *price range* of materials. In order to satisfy this need, a chart has been prepared that shows a scale for the relative position of each material with respect to the others in terms of cost per cubic inch (Fig. 5−1). This scale is in numbers that are multipliers of the "base unit." The base unit is derived from the average cost per cubic inch of four utility materials, namely, polystyrene, phenolic, polyethylene, and polypropylene. This step becomes necessary owing to constant price fluctuation. It was established over a period of several years that the average cost of the four materials was in a reasonable constant relationship to the prices of other popular materials.

As an example, if the chart was used in May of 1981, the base unit would be

Polystyrene	1.9 ¢/in.^{3}
Phenolic	2.9 ¢/in.^{3}
Polyethylene	1.45 ¢/in.^{3}
Polypropylene	1.45 ¢/in.^{3}
Total	7.70 ¢/in.^{3}

Average ¢/in.^{3} or "base unit" = 1.925.

Thus the price of polycarbonate, reading on the chart 3.43, would be $3.43 \times 1.925 = 6.6$ ¢/in.^{3} or for nylon 66, the reading is $3.48 \times 1.925 = 6.7$ ¢/in.^{3}. During any time period, the base unit is updated by obtaining the prevailing price of the four materials in terms of cents per cubic inch.

Occasionally, some material may fall out of the range shown on the chart, in which case a chart correction should be made.

The plain polymer price units on the chart apply to general purpose or popular grades of material.

It should be recognized that modified polymers that will meet specific requirements, such as high impact strength, higher heat resistance, UV stability, and flame retardance, will be more expensive, and before any material is adopted for a product, it should be checked for prevailing costs. For purposes of getting a general idea about the group of materials to be considered for a contemplated product, the following procedure is suggested.

Most products are designed to meet proposed cost limit. In this event it can be estimated roughly that the material cost will constitute about 40%−70% of the total cost. The lower percentage would cover utility materials, whereas the higher percentage would include nylons and polycarbonates and similar polymer categories.

It is assumed that a preliminary product sketch is available which would make it possible to estimate its volume. These cubic inches divided into the total material cost will give a price per cubic inch that will, from the chart, indicate several materials for consideration.

A more accurate way to point to a material group is to estimate the manufacturing cost of the product as outlined in Chapter 10, subtract this cost from the total cost limit, thereby obtaining the material value of the product. Here again, as in the previous case, the volume of the part is divided into the material cost and figures are obtained that will lead to a particular plastic on the chart.

For example, a product with a cost limit of $1.00 has a material content of 60¢ and a volume of 10 cubic inches: *60¢* divided by 10 cubic inches gives a cost of 6 ¢/in.^{3}, which leads to nylon, acetal, polycarbonate, and others on the chart.

The final material selection method is outlined in Chapter 7, but the above method is merely to make a rough first guess.

The chart shows a scale for "Units—low" and refers to the lower- to medium-priced material range. The scale marked "Units—high" points to high-priced "exotic" materials.

The two groups are separated by the sectioning symbols to indicate the prices for 10, 20, 30, and 40% glass-containing materials. Materials that vary in resin content, as well as amount and type of filler in each grade, are impractical to show on a chart of this type. The materials in this category are thermoset polyesters and PVC compounds. The price per pound of the thermoset polyester resin alone is close to the price per pound of the phenolic molding compound. The price of the PVC resin alone is nearly comparable to that of polyethylene. Both are on a per pound basis.

It is to be reemphasized that this chart is based on units which represent cost per cubic inch of popular material grades and is intended to provide a guide to the designer to choose a tentative group of plastics that should be investigated for product application. This step is to be followed by specific grade selection and price calculations that will meet design criteria.

6. Product Design Features

One of the earliest steps in product design is to establish the configuration of the parts that will form the basis on which strength calculations are made and a suitable material selected for anticipated requirements.

During the sketching and drawing of shape and cross sections, there are certain design features with plastic materials that have to be kept in mind in order to avoid degradation of test data properties. Such features may be called property detractors, since most of them are responsible for internal stresses that reduce the available stress level for load-bearing purposes. Other features that are covered in this chapter may be classified as precautionary measures that may influence the favorable performance of a part if properly incorporated.

Let us review these property detractors along with other features and suggest means of circumventing their potential negative effects.

INSIDE SHARP CORNERS

Sharp corners on the inside of a part are the most frequent property detractors. When the part drawing does not show a radius, the tendency is for the toolmaker (while making a mold) to leave the intersecting machined or ground surfaces as generated by the machine tool, the result being a sharp corner on the molded part.

The material data sheet shows the difference in impact strength between notched and unnotched test bars. In some materials this ratio is 1 to 30, and in others there is also a decided reduction in strength.

In a shaped product, the inside sharp corner means that a specified tough material acts in a brittle manner. Sharp corners are stress concentrators.

The stress concentration factor increases as the ratio of the radius R to the part thickness T decreases. An R/T of 0.6 is favorable, and an increase in this value will be of limited benefit. There are other details in this problem that when properly counteracted will help in reducing stress concentration. In Fig. 6−1 we see that a concentric radius, in addition to eliminating the outside sharp corner, can play an important part in holding down the value of the stress concentration.

The Izod impact strength of nylon with various notch radii is shown in Fig. 6−2. Thus we see that with an 0.005R the impact strength is about 1.3 ft-lb/in.; with 0.020R, it is 4.5 ft-lb/in.; and with 0.040R, it is 12 ft-lb/in.

In most cases an 0.020R can be considered a sharp corner as far as end use is concerned, and this size is a decided improvement over 0−5-mil radius; therefore, it should be considered as a minimum requirement and should be so specified. If this radius of 0.020R would cause interference, a corner as shown in Fig. 6−3. should be considered.

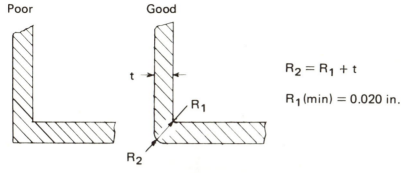

Poor   Good

$$R_2 = R_1 + t$$

$$R_1 (\text{min}) = 0.020 \text{ in.}$$

Fig. 6–1. An inside corner.

The recommended radius not only reduces the brittleness effect, but also provides a streamlined flow path for the plastic in the mold. The radiused corner of the metal in the mold reduces the possibility of its breakdown and thus eliminates a potential repair need.

Too large a radius is also undesirable since it wastes material, may cause sink marks, and even contributes to stresses due to excessive variation in thickness.

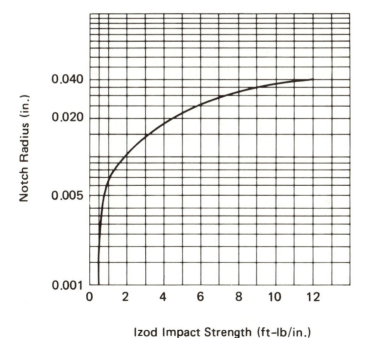

Fig. 6–2. Radius of notch versus notched Izod impact strength.

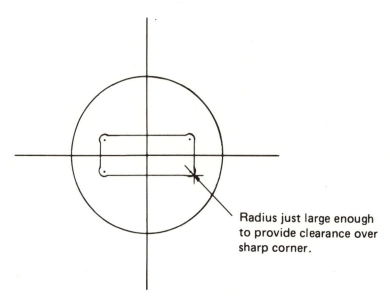

Radius just large enough
to provide clearance over
sharp corner.

Fig. 6–3. Clearance radius for sharp corner.

In summary, the designer should be aware of the high degree of notch sensitivity in plastics, and thus eliminate inside sharp corners and indicate an appropriate radius.

UNIFORM WALL THICKNESS

Wall requirements are usually governed by load, support needs for other components, attachment bosses, and other protruding sections.

Designing a part so that all these requirements will end up in a reasonably uniform wall will greatly benefit its durability. A uniform wall thickness will minimize stresses, differences in shrinkage, possible void formation, and sinks on the surface; it also usually contributes to material saving and economy of production.

Most of the features for which heavy sections are intended can be modified by means of ribbing, coring, and shaping of the cross section to provide equivalent strength, rigidity, and performance. Figure 6–4 shows a small gear manufactured from metal bar stock. The same gear converted to a molded plastic would be designed as shown in Fig. 6–5. The plastic gear design, as compared with a copy of the metal gear, saves material, eliminates stresses due to a thick and thin section, shrinkage in teeth and the remainder of the gear is the same, avoids the danger of warpage, thin web and tooth base prevents bubble formation and potential weak spots, and sink in the middle of the thickness is absent thus ensuring full load carrying capacity of the teeth.

Figure 6–6 illustrates other cross sections that are either well or poorly designed.

t = Thickness through
pitch line

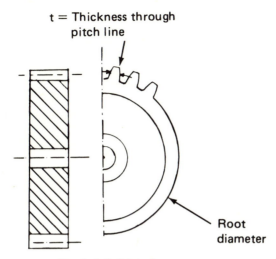

Fig. 6-4. Solid steel gear.

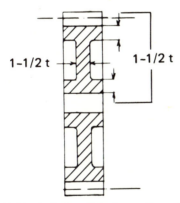

Fig. 6-5. Plastic design of steel gear shown in Fig. 6-4.

If, however, a case exists where some thickness variation is unavoidable, then the transition should be gradual to prevent sharp changes in temperature during solidification.

RIBS

In the discussion of uniform wall thicknesses, ribbing was one of the suggested remedies. Ribs are also used to increase load bearing requirements when calculations indicate wall thicknesses are above recommended values. They are provided for spacing purposes, for supporting components, etc.

The first step in designing a rib is to determine dimensional limitations followed by establishing what shape the rib is to have in order to realize a part

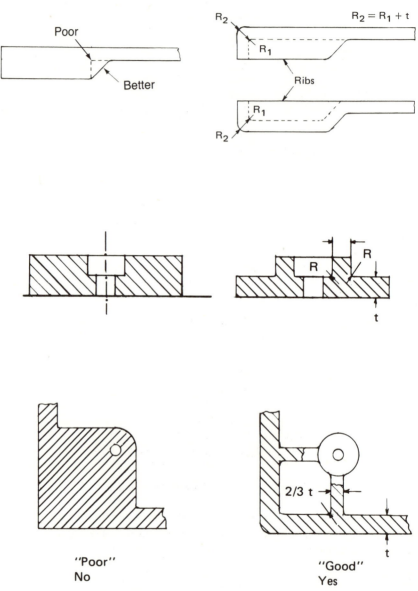

Fig. 6-6. "Poor" or problem and "good" or successful design features are shown.

with good strength and satisfactory appearance that can be produced economically. Figure 6-7 shows proportional dimensions of rib versus thickness. This arrangement will minimize voids (sinks), stresses, and shrinkage variations and lends itself to trouble-free molding.

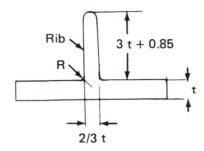

R = 0.030 to 0.050 in.

Fig. 6–7. Rib proportion versus thickness.

If performance calculations indicate wall thicknesses well above those recommended for a particular material, one of the solutions to the problem is to find equivalent cross-sectional properties by ribbing. Heavy walls are responsible for reduction in properties due to poor heat conductivity during molding, thus creating temperature gradients throughout the cross section, thereby causing stresses. Cycle times are usually exceptionally long, thus adding another cause for stresses. Also, close tolerance dimensions are difficult to maintain, material is wasted, quality is degraded, and cost is increased. Solid plastic wall thicknesses for most materials should be below 0.2″ and preferably around 0.125″ in the interest of avoiding the above pitfalls. In most cases ribbing will provide a satisfactory solution; in other cases reinforced material may have to be considered.

An example of how ribbing will provide the necessary equivalent moment of inertia and section modulus will now be given.

A flat plastic bar of 1½″ × ⅜″ thick and 10″ long, supported at both ends and loaded at the center, was calculated to provide a specified deflection and stress level under a given load. The favorable material thickness of this plastic is 0.150, and rib proportions would be per Fig. 6–7.

Using judgment as a guide, it would appear that the 1½″ width would require about two ribs. So, as a starting point, we will calculate the equivalent cross-sectional data as if we were dealing with two "T" sections.

According to the handbooks under "Stress and Deflections in Beams" and "Moments of Inertia," etc., the resistance to stress is expressed by the moment of inertia and resistance to deflection by the section modulus. By finding a cross section with the two equivalent factors, we will ensure equal or better performance.

The stress $= Wl/4Z$ and deflection $= Wl^2/48EI$, where W = load, l = length of beam, Z = section modulus, E = modulus of elasticity, and I = moment of inertia.

For the flat bar $I = bd^3/12$, $b = 1½$, $d = ⅜$; or

$I = (1.5 \times 0.375^3)/12 = 0.0066$

for the rectangular bar. The section modulus is

$$Z = \frac{I}{Y} = \frac{0.0066}{0.1875} = 0.0352$$

The moment of inertia of a "T" section is

$I = \frac{1}{12}[4bs^3 + h^3(3t + T)] - A(d - y - s)^2$

where $b = 0.75$, $s = 0.15$, $h = 0.6$, $t = 0.08$, $T = 0.1$, $d = 0.75$, $A =$ area $= bs + h(T + t)/2$.

$y = d - [3bs^2 + 3ht(d + s) + l(T - t)(h + 3s)]/6A$

where y is the distance from the neutral point to the extreme fiber. Substituting the values into the formulas we have:

$$A = (0.75 \times 0.15) + \frac{0.6(0.1 + 0.08)}{2} = 0.1665$$

$$y = 0.75 - \{(3 \times 0.75 \times 0.15^2) + [3 \times 0.6 \times 0.08(0.75 + 0.15)] \\ + 0.6(0.10 - 0.08)(0.6 + 0.45)\}/(6 \times 0.1665)$$

$$= 0.75 - \frac{0.0506 + 0.1296 + 0.0126}{0.999} = 0.75 - 0.193 = 0.557$$

$$I = \frac{1}{12}[(4 \times 0.75 \times 0.15^3) + 0.6^3\{(3 \times 0.08) + 0.1\}] \\ - 0.1665(0.75 - 0.557 - 0.15)^2$$

$$= \frac{1}{12}[0.0102 + 0.073] - 0.1665 \times 0.00185$$

$$= 0.00693 - 0.00031 = 0.00662$$

$$Z = \frac{0.00662}{0.557} = 0.0119$$

Two of the "T" sections would provide a higher moment of inertia and decreased section modulus. When placed on the end, the two ribs would make a channel that would give a moment of inertia of 0.018 and a section modulus of 0.035, which values are more than adequate for the purpose. (See Fig. 6–8.)

It should be noted that the two-rib construction forming a channel would use 70% of the material compared to a solid bar. (See Fig. 6–9.)

Other means of stiffening surfaces can also be employed—if appearance permits—where areas can be domed or corrugated. The basic goal in any action that leads to greater rigidity is to specify practical wall thickness that will optimize strength and processing, and thus result in high-quality products.

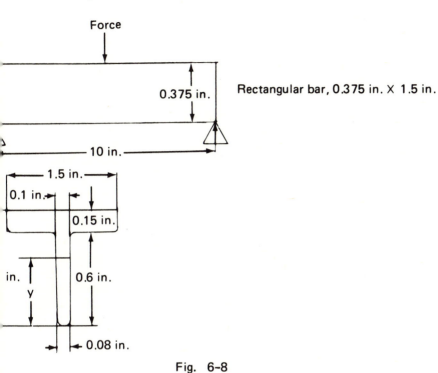

Fig. 6-8

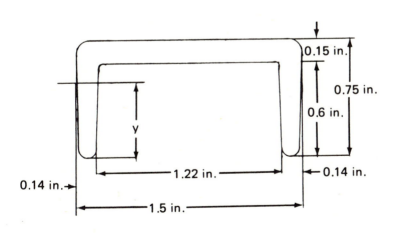

Fig. 6-9

Figs.6−8 and 6−9. Shapes with good moldability that give a section modulus and moment of inertia that are equivalent to a rectangular bar.

In addition to ribs, other protruding features from a wall, such as bosses or tubular shapes, should be treated similarly as far as transition radius, taper, and minimal material usage are concerned. The same principles are involved and identical ill effects can be expected unless recommended practices are incorporated.

TAPERS OR DRAFT ANGLES

It should be recognized that any vertical wall of a molded product should have an amount of draft that would permit easy removal from a mold. Figures 6–10 and 6–11 show two basic conditions in which draft is a consideration.

Figure 6–10 is the most desirable application of the draft angle and the amount of draft may vary from ⅛° up to several degrees depending on what the circumstance will permit. A fair average may be from ½° to 1°. When a small angle such as ⅛° is used, the outside surface—the mold surface producing it—will require a very high directional finish in order to facilitate removal from the mold.

The other case, shown in Fig. 6–11, may represent a separating inside wall which generally should be perpendicular to the base. The draft in this case should be on the low side ⅛° so that additional material usage is small, the

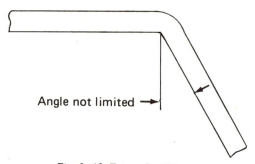

Angle not limited ➞

Fig. 6–10. External wall taper.

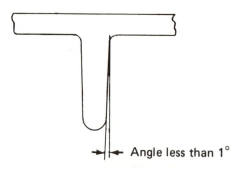

➞|➞ Angle less than 1°

Fig. 6–11. Internal wall taper.

possibility of voids close to the base is avoided, and increased cycle time in manufacture is minimized. Here again the vertical molding surfaces will demand a much higher surface finish with polishing lines in the direction of part withdrawal.

On shallow walls, the draft angle can be considerably larger, since the influence of the enumerated drawbacks will be minor in their effect. The designer should be cognizant of the need for drafts on vertical walls. If problems are encountered during removal of parts, stresses can result, the shape of the product can be distorted, and surface imperfections can be introduced.

Other features encountered in product design call for careful consideration in order to avoid potential ill effects on product performance. The following paragraphs will enumerate the most essential features that require attention.

1. Weld lines (Fig. 6−12). When a part having openings (holes) is formed, such openings are generated by cores. In the process of filling a cavity, the flowing plastic is obstructed by the core, splits its stream, and surrounds the core; the split stream then reunites and continues flowing until the cavity is filled. The rejoining of the split stream forms a weld line that does not have the strength properties that an area without a weld line would have. The reason for the reduction in strength is that the flowing material tends to wipe air, moisture, and lubricant into the area where the joining of the stream takes place and introduces a foreign substance into the welding surface. Furthermore, since the plastic material has lost some of the heat, the temperature for self-welding is not conducive for most favorable results. A surface that is subjected to load bearing should not contain weld lines; if this is not possible, the allowable working stress should be reduced by at least 15%.

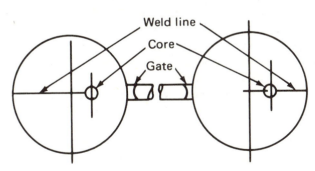

Fig. 6−12. Locating the weld line to suit load.

2. Gate size and location. The area near a gate is highly stressed due to frictional heat generated at the gate and the high velocities of the flowing material. A small gate is desirable for separating the part from the feed line, but is not desirable for a part with low stresses. Gates are usually two-thirds of the part's thickness, and if they are of such dimensions or larger will reduce frictional heat, permit lower velocities, and allow the application of higher pressures for densifying the material in the cavity. The product designer should alert the tool designer to keep the gate area away from load bearing surfaces and to make the gate size such that it will improve the quality of a product.

3. *Wall thickness tolerance*. When relatively deep parts such as instrument covers and similar pieces are designed, a tolerance of the wall thickness on the order of $\pm 0.005''$ is usually given. What this tolerance should mean is that a product will be acceptable when made with this tolerance, but that the wall thickness must be uniform throughout the circumference (see Fig. 6–13).

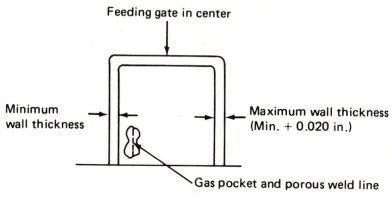

Figure 6–13

If we analyze the molding conditions of the part in Fig. 6–13 and assume that one side is made to the minimum of specifications and the opposite to the maximum, we find the following taking place: The resistance to the plastic flow decreases with the third power of the thickness, which means that the thick side will be filled first, while the thin side is filled from all sides instead of the gate side alone. This type of filling creates a pocket on the thin side and compresses the air and gases to such a point that the rising temperature due to compression causes the material to be charred while the pocket is being filled up. This charred material will create porosity, a weak area, and an electrically defective surface. Furthermore, the filling of the thick side ahead of the thin side creates a pressure imbalance generated by the 5–10 tons per square inch injection pressure that can cause the core to deflect toward the thin side and further aggravates the difference in wall thickness. This pressure imbalance will contribute to mold damage and make part production difficult if not impossible.

The conclusion from this explanation is that the *wall uniformity* throughout the circumference must be within narrow limits such as $\pm 0.002''$, whereas the thickness in general may vary within the specified $\pm 0.005''$.

4. *Molded-in inserts*. If metal inserts are to be molded into a product, their shape should present no sharp edges to the plastic, since the effect of the edges would be similar to a notch. A knurled insert should have the sharp point smoothed, again to avoid the notch effect (see Fig. 6–14). Let us look at the

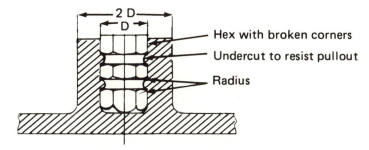

Fig. 6-14. Preferred molded-in insert.

merits of molding inserts in place. This procedure is usually employed to provide good holding power for plastic products. There are certain drawbacks to this method: it is dangerous to have an operator place an arm between the mold halves while the electrical power to the machine is "ON"; it normally takes a pin to support the insert, and since this pin is small in relation to the cored hole for the insert, it is easily bent or sheared under the influence of injection pressure; should the insert fall out of position, there is danger of mold damage; the hand placement of inserts contributes to cycle variation and with it, potentially, product quality degradation. Some of these problems can be overcome by higher mold expenditures, for example, shuttling cavities.

On the other hand, desired results in fastening can be attained by other means, for example, (a) coring holes in the part that will permit ultrasonic welding of inserts in place; (b) coring a hole in the part that will be of a size when the part is removed from the mold that will permit a slight press fit plus a gain in the holding power from postmolding shrinkage (Fig. 6-15); (c) coring a hole in the part that will permit dropping the insert and providing a retaining shoulder by spinning or ultrasonic forming (Fig. 6-16).

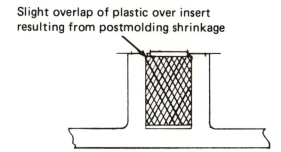

Fig. 6-15. Light press fit of insert plus shrinkage.

All these assembly methods require the same time to perform as placing inserts in the mold, but, in addition, they also lower machine time.

Before spinning or
welding

After spinning or welding

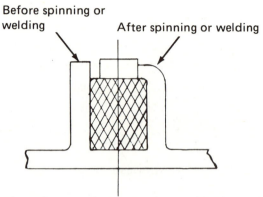

Fig. 6–16. Insert dropped into cored opening and retained by closing plastic over metal shoulder.

There are probably several other means of accomplishing the desired result that depend on the circumstances at hand. In any event, molded-in inserts, in the long run, prove costlier, therefore, they should be avoided.

In the electronic field there are molded-in stampings or wire inserts with an activity of 10,000 or more parts per run. In these cases, it usually involves a combination of stamping or wire-forming equipment with that of molding in which the placement of the insert is automatically performed.

5. *Internal plastic threads*. The strength of plastic threads is limited, and when molded in a part involving either an unscrewing device or a rounded shape of a thread—similar to bottle cap threads—they can be stripped from the core. Screw threads, when needed, should be of the coarse type and have the outside of the thread rounded so as not to present a sharp V to the plastic, which can produce a notch effect (Fig. 6–17).

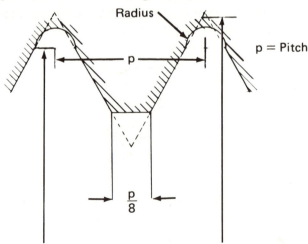

Fig. 6–17. Inside thread in plastics.

If a self-threading screw can be substituted, it would not only decrease appreciably the mold maintenance and mold cost, but, most likely, with proper type selection, will give a better holding power. A screw that has a thin thread with relatively deep flights can give high holding power. If the screw or plastic is preheated to about 250°F (121°C), a condition of thermoforming in combination with material displacement will exist, thereby resulting in more favorable holding power. When male plastic threads are being considered, the coarser threads are again preferred and the root of the thread should be rounded to prevent the notch effect.

6. Blind holes. The pressures exerted by the flowing plastic in injection molding are high. A core pin forming blind holes is subjected to the bending forces that exist in the cavity. Calculations can be made for each case by establishing the core pin diameter, its length, and anticipated pressure conditions in the cavity. From technical handbooks, we know that a pin supported on one end only will deflect 48 times as much as one supported on both ends. This suggests that the depth of hole in relation to diameter be small in order to maintain a straight hole. There are cases where a deep and small diameter hole is needed, as in the example of pen and pencil bodies. In this case the plastic flow is arranged to hit the free end of the core from four to six evenly spaced gates that will cause a centering action, and the plastic will continue flowing over the diameter in an umbrella-like pattern to balance the pressure forces on the core. Where this type of flow pattern is impractical, an alternative may be a through hole (or tube formation) combined with a postmolding sealing or closing operation by spinning or ultrasonic welding.

In the other extreme, let us consider a $\frac{1}{4}''$ diameter core, exposed to a pressure of 4000 psi with an allowance of deflection of $0.0001''$ and see how deep a blind hole can be molded under those conditions.

According to handbooks the deflection is

$$a = \frac{Wl^3}{8EI} = \frac{1000l^4}{8 \times 30,000,000 \times 0.049 \times 0.0039}$$

$$I = \frac{\pi d^4}{64} = 0.049d^4$$

$$d^4 = 0.0625 \times 0.0625 = 0.0039$$

where W = total load = psi × d × l (projected area of pin) = 4000 × $\frac{1}{4}$ × l = $1000l$: l = length of pin; d = diameter of pin; a = deflection = 1/10,000,

$$\frac{1}{10,000} = a = \frac{1000l^4}{8 \times 30,000,000 \times 0.049 \times 0.0039} = \text{deflection}$$

$$l^4 = 0.0045864$$

$$l = 0.26 \text{ or slightly over diameter size.}$$

If a hole deeper than 0.26″ is needed, we can calculate the amount of deflection that will be present and whether the calculated deflection will produce an opening of the necessary tolerance and the kind of stress that will be generated in the pin along with its corresponding life expectancy. Let us now assume that the desired depth of hole is ⅜″.

The deflection

$$a = \frac{Wl^3}{8EI} = \frac{1000l^4}{8 \times 30,000,000 \times 0.049 \times 0.0039}$$

$$l^4 = 0.0198$$

$$a = \frac{0.0198}{45.864} = 0.00043'' \text{ deflection}$$

The maximum stress

$$S = \frac{Wl}{2Z} = \frac{1000l^2}{2 \times 0.006125} = \frac{1000 \times 0.1406}{2 \times 0.006125}$$

$$Z = 0.098d^2 = 0.006125$$

$$S = 11,480 \text{ psi}$$

These results indicate that a hole with 0.0004″ variation may be satisfactory, and if the pin is made of a springlike material, properly heat treated, it should last a long time.

7. Undercuts. Whether the undercuts are external or internal, they should be avoided if possible. In cases where it is essential to incorporate them in part design, a great many of them can be realized by appropriate mold design in which either sliding components on tapered surfaces or split cavity cam actions will produce a needed undercut. This obviously means an increased tool cost in the neighborhood of 15–30%.

There are, however, some conditions that will permit incorporating undercuts with conventional stripping of the part from mold. Certain precautions are necessary in order to attain satisfactory results. First, the protruding depth of undercut should be two-thirds of wall thickness or less. Second, the edge of the mold against which the part is ejected should be radiused to prevent shearing action, and, finally, the part being removed should be hot enough to permit easy stretching and return to the original shape after removal from mold (Fig. 6–18).

How easily the task can be accomplished depends on material elasticity and spring back. Many threaded plastic caps are stripped from the cores instead of being unscrewed. Coarse threads with the crest of the core thread rounded and a material with good elongation and ability to spring back make it feasible to apply conventional part stripping. The undercut problems can be solved by the

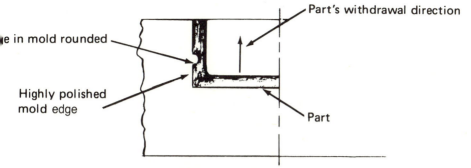

Fig. 6–18. External undercut.

cooperation of designer, mold maker and processor, since each product configuration presents different possibilities.

8. Thermoplastic hinge. Hinge dimensions of lids and boxes made of thermoplastics like polypropylene have been well established and are shown in Fig. 6–19. The successful operation of such a hinge depends not only on design, but also on ensuring that the flow pattern of the plastic through the hinge stretches the molecules to give a strong and pliable hinging section. In larger sized parts, there is a tendency to place a gate at the center of the box and another at the center of the lid with the result that the flow pattern is not conducive to creating favorable hinge strength.

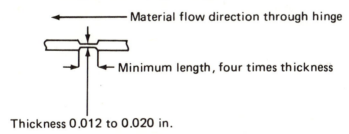

Fig. 6–19. Thermoplastic hinge.

9. Functional surfaces and lettering. Surfaces of plastics may be provided with designs that can give a good grip or that can simulate wood, leather, etc. These types of surfaces should be specified in a manner that will not create undercuts to the withdrawal action from a mold. The undercut effect can be responsible for stresses and marring. A similar condition applies to lettering, and the location of such lettering should conform to smooth withdrawal requirements.

10. Tolerances. There is an industry standard for specifying limits for certain dimensions, and each material supplier converts the data to suit a specific material. Figure 6–20 shows the information for polycarbonate. This type of information is intended to give the designer a guide for tolerances that

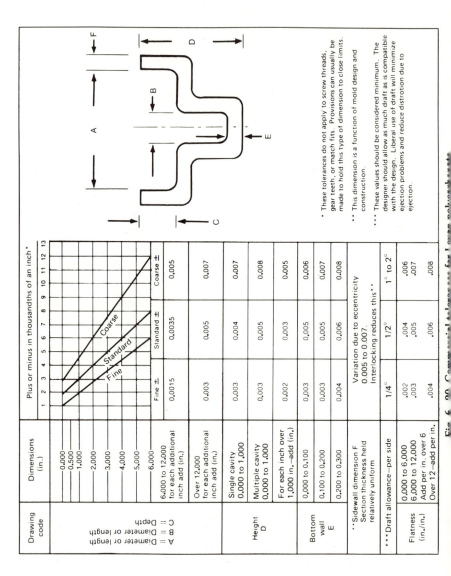

Fig. 6-20 Commercial tolerances ... for Lexan polycarbonate

are to be shown on the drawing; these tolerances would include variation in part manufacture and some variation in tooling.

Tolerances on dimensions should be specified only where absolutely necessary. Too many drawings show limits of sizes where other means of attaining desired results would be more constructive. For example, if the outside dimensions of drill housing halves were to have a tolerance of ±0.003", this would be a tight limit. And yet if half of the housing were to be on the minimum side and the other on the maximum side, there would be a resulting step that would be uncomfortable to the feel of the hand while gripping the drill.

A realistic specification would call for matching of halves that would provide a smooth joint between them, and the highest step should not exceed 0.002". The point is that limits should be specified in a way that those responsible for the manufacture of a product will understand the goal that is to be attained. Thus we may indicate "dimensions for gear centers," "holes as bearing openings for shafts," "guides for cams," etc. This type of designation would alert a mold maker as well as molder to the significance of the tolerances in some areas and the need for matching parts in other places and clearance for assembly in still other locations.

Most of the engineering plastics reproduce faithfully the mold configuration, and when processing parameters are appropriately controlled, they will repeat with excellent accuracy.

We see plastic gears and other precision parts made of acetal, nylon, polycarbonate, and Noryl, whose tooth contour and other precision areas are made with a limit of 0.0002", and the spacing of the teeth is uniform to meet the most exacting requirements.

The problem with any precise part is to recognize what steps are needed to reach the objective and follow through in a detailed manner every phase of the process to safeguard the end product.

In Chapter 2, under "shrinkage," the factors that can cause variation in shrinkage are listed, and from that alone, we see how processing parameters can influence dimensional variation. Some materials perform better than others in that respect.

Generally speaking, if we segregate the tolerances into (a) functional need, such as running fit, sliding fit, gear tooth contour, etc.; (b) assembly requirements that are to accommodate parts with their own tolerances; (c) matching parts for appearance or utility; we should come up with feasible tolerances that will be reasonable and useful. This approach will be more productive than trying to apply tolerances strictly on a dimensional basis.

Tolerances should be indicated only where needed, carefully analyzed for their magnitude, and proven out as to their usefulness.

Adaptation of metal tolerances to plastics is not advisable. With plastics reaction to moisture and heat, for example, is drastically different from metals, so that pilot testing under extreme use conditions is almost mandatory for establishing adequate tolerance requirements.

7. Designing the Plastic Product

When reading the literature or magazines in which the design of products with plastics is discussed, there is an impression that special rules are involved in dealing with this material. A more desirable approach to the subject is to look at the project as just another design where a different material is employed. With this attitude, the problem analysis will be as usual, the formulas the same as for other materials, but the insertion of values for properties will require abnormal care.

Technical handbooks with all the formulas and other descriptive explanations are just as much a necessity in this case as with other older and well-established materials.

To do full justice to the plastic material, we should become familiar with the meaning of the data furnished by the material supplier, the long-term effect on properties when subjected to a load, the influence of surrounding environment, and other material requirements that, as a rule, may be insignificant with metals. These subjects are fully described in Chapters 2, 3, 5, and 6.

With the information covered in these chapters clearly in mind, it will become rather routine to follow through with a design, to build a prototype, and to test performance capability.

The design examples used in this chapter were selected for the purpose of demonstrating that the standard technical handbooks are in general of the same importance in conjunction with plastics as they are with other materials.

On the basis of information from the preceding chapter, the part sketches or drawings have been modified to eliminate strength detractors and to incorporate needed features in such a manner that overall part characteristics will be protected. The next step in the program is to specify the appropriate material. Before investigating any one or a group of materials, a number of questions have to be answered as to the end use performance and environmental conditions. Some of these questions are:

1. What are conditions under which the product will be used: Temperature, moisture, ultraviolet exposure, exposure to fungus, flammability, chemicals, electrical resistance, arc resistance, light transmission, stability or permanency, physical properties, colorability, heat insulation, resistance to scratching (mar resistance), and special requirements such as self-lubrication, lightness, hinging property, spring properties, time of exposure, etc.?

2. What tolerance requirements are expected in the performance of the product? Do shrinkage characteristics of the contemplated material appear small so that tolerances can be anticipated with a reasonable degree of accuracy?

3. What is the nature of the load to which the product will be exposed, such as impact, creep, deflection, stresses, bending, gliding, etc.?

4. Is color matching a factor?

5. What is the cost per cubic inch of material? Its weight?

6. Approval requirements: Underwriters', Pure Food Administration, military, etc.

Once the performance and environmental conditions have been defined, the selection of a suitable material can be made, and this in turn can be followed with the necessary calculations to establish cross-sectional strength requirements. The basic data needed for calculations have to be collected and have to pertain to the specific grade of the selected material. The pertinent information required for making determinations for longevity of the product and obtaining a general concept of the character and behavior of the selected material should be supplied by the manufacturer of the raw material and should be along the following lines:

1. Data sheets of the specific grade of material, which are to contain all the properties listed in data chart, Fig. 7−1.

2. Stress−strain curves at the conditions of product application. This would indicate the toughness of material by sizing up the area under the curve. It would also show the proportional limit, yield point, and corresponding elongations.

3. Curves showing change of tensile strength, flexural strength, and modulus with increasing temperatures.

4. Creep data for periods at 100 and 1000 hours (or more, if available) covering stress and temperature conditions closely comparable to those of product application.

5. The allowable working stress, based on successful performance at conditions of product usage.

6. Chemical resistance at conditions of application, if needed.

7. A statement of major applications comparable in design parameters to those of the contemplated product.

Not all product components are subjected to a load. In those cases, the designer is concerned with features that will prevent internal stresses, with elements that will lead to consistent and economical production, and with appearance and dimensional control.

Parts that are subjected to a load have to be analyzed carefully with respect to the duration of the load, the temperature conditions under which the load will be active, and the stress created by the load.

A load can be defined as continuous when it remains constant for a period of 6 hours, whereas an intermittent load is considered one that is of 2 hours duration and is followed by an equal time for stress recovery.

Category	PROPERTY	ASTM TEST METHOD
PHYSICAL	Mold shrinkage (in./in.)--in mils	D955
	Specific gravity	D792
	Specific volume (in.^{3}/lb)	
	Water absorption (%) 24 hr @ 73F (23C)	D570
	equilibrium, 73F (23C)	
	Haze (%)	
	Transmittance (%)	D1003
MECHANICAL	Tensile strength (psi) yield	D638
	ultimate	
	Elongation (%) yield	D638
	rupture	
	Tensile modulus (x10^5 psi)	D638
	Flexural strength (psi)	D790
	Flexural modulus (x10^5 psi)	
	Compressive strength (psi)	D695
	Shear strength (psi)	D732
	Izod impact strength notched, 1/8 in.thick	D256
	(ft-lb/in.) unnotched, 1/8 in.thick	
	Tensile impact strength (ft-lb/in.^{2}) S type	D1822
	L type	
	Fatigue strength (psi @ ___ cycles)	D671
	Rockwell hardness	D785
	Deformation under load (%) @ 73F (23C) @ ___ @ ___ F	D621
THERMAL	Heat deflection temperature (F) @ 66 psi	D648
	@ 264 psi	
	Specific heat (Btu/lb/F or Btu/lb/C)	
	Thermal conductivity (Btu/hr/ft^2/F/in.)	
	Coeff. therm. exp. (x10^5) (in./in./F or /C)	D696
	Brittleness temperature (F)	D746
	Flammability UL Standard 94	UL94
	Oxygen index	D2863
ELECTRICAL	Dielectric strength (V/mil)	D149
	Dielectric constant 60H	D150
	10^{6}H	
	Power factor 60H	D150
	10^{6}H	
	Volume resistivity (ohm-cm)	D150
	Arc resistance (sec), tungsten electrodes	D495

The temperature factor requires greater attention than would be the case with metals. The useful range of temperatures for plastic applications is relatively low and is of a magnitude that in metals is viewed as negligible.

Now we come to an allowable working stress for a specific material. The viscoelastic nature of the material requires not merely the use of data sheet information for calculation purposes, but also the actual long-term performance experience gained that can be used as a guide. The allowable working stress is important for determining dimensions of the stressed area and also for predicting the amount of distortion and strength deterioration that will take place over the life-span of the product. This means that the allowable working stress for a constantly loaded part that is expected to perform satisfactorily over many years has to be established using creep characteristics of a material that has sufficient data with which a reliable long-term prediction of short-term test results can be made.

We will now analyze creep data and determine guidelines on how they should be used. As stated elsewhere in this volume, the viscoelastic nature of plastic material reacts to a constant load over a long period of time by an ever increasing strain. Since the modulus formula states that modulus = stress/strain and the stress is constant, while the strain is increasing, we see an ever decreasing modulus. This type of modulus is called an *apparent modulus*, and the data for it are collected from test observations for the purpose of predicting long-term behavior of plastics subjected to a constant stress at selected temperatures.

There have been attempts made to create formulas for the apparent modulus change with respect to time, but the factors in the formulas that would fit all conditions are more cumbersome to use than presenting test data in a graph form and using it as the means for predicting the strain (elongation) at some distant point in time. The test data when plotted on log-log paper usually form a straight line and lend themselves to extrapolation. The slope of the straight line indicating a decreasing modulus depends on the nature of a material (principally its rigidity and temperature of heat deflection), on the temperature of the environment in which the part is used, and on the amount of stress in relation to tensile strength.

The author has plotted most of the available creep data (amounting to over 600 charts) test results from which certain conclusions were drawn.

1. For practical design purposes, the data accumulated up to 100 hours of creep is of no real benefit. There is usually too much variation during this test period, which is of relatively short duration.

2. The apparent modulus values starting with a test period of 100 hours up to 1000 hours when plotted form a straight line on log-log graph paper.

3. This line may be continued for longer periods on the same slope for interpolation purposes, provided the stress level is one-quarter to one-fifth that of the

ultimate tensile strength and the test temperature is no greater than two-thirds of the difference between room temperature and the heat deflection temperature at 264 psi. This conclusion was verified by plotting the available creep data for time periods greater than 1000 hours.

When the limitations outlined above are exceeded, there is a sharp decrease in apparent modulus after 1000 hours with indications that failure due to creep is approaching, i.e., the material has attained the limit of its usefulness.

4. Since the designer will be expected to plot curves to suit his requirements, some examples will be cited that can serve as a guide for potential needs. (See Fig. 7–2.)

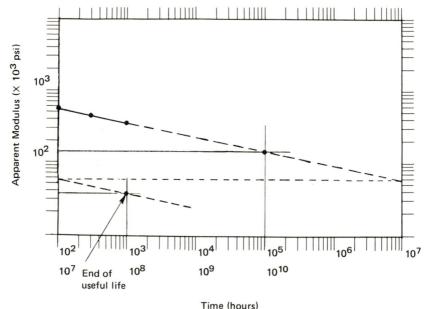

Fig. 7–2. Creep Lines for ABS.

We will now examine the ABS creep line for 1000 psi stress at 73°F (23°C) temperature. When the line is extended to 10^5 hours, we find an apparent modulus of 140,000 psi. If the product is designed for the duration of 10^5 hours and calculations are made for part dimensions, the modulus of 140,000 psi should be inserted into any formula in which the modulus appears as a factor.

At 10^5 hours we will have a total strain of

$$E = \frac{\text{stress}}{\text{strain}}$$

$$140,000 = \frac{1000}{\text{strain}}$$

$$\text{strain} = \frac{1000}{140,000} = 0.007 \text{ or } 0.7\%$$

If the product can tolerate this type of a strain without affecting its performance, then the dimensional requirements are met.

The elongation at yield for ABS is 0.0275, which could be considered the end of the useful strength of the material. Let us see what apparent modulus would correspond to this strain at 1000 psi and 73°F (23°C) and at what time it would be reached.

$$E = \frac{1000}{0.0275} = 36,364 \text{ psi}$$

Using the lower part of the graph, we draw at the point of 56×10^7 on the left side a line parallel to the original creep line and find that it intersects the time of $10^8 \times 1$ hours. This means the product would fail at that time owing to the loss of strength even if dimensional changes would permit satisfactory function of the product.

Other charts show creep test data beyond the 1000-hour duration; under most conditions the straight line between the 100- and 1000-hour points is continued into the 10,000- and 20,000-hour range. Even in these charts, we occasionally find a deviation from the straight line, which should not be considered unreasonable because of all the variables that enter into the test data. We must recognize that variation in material, in preparation of the specimen, in test conditions, and in testing devices, and possible human errors, can be factors in the results on which the graphs are based.

Some variations on the charts only amount to a few percent, other, larger discrepancies are due to high temperatures and stress levels that are beyond the limits suggested in the early paragraphs of this discussion.

The criteria for selecting an allowable continuous working stress at the required temperature have to be such that it is possible to make an interpolation of the elongation at the end of the product life. For example, if a product will be stressed to 1700 psi at a temperature of 150°F (66°C) and data are available for 2000 psi stress at 160°F (71°C), this information plotted on log-log paper should provide the possibility of extrapolating the long-term behavior of the material.

The intermittent stress can usually be twice that of the continuous stress.

Most suppliers conduct creep tests up to 1000 hours with the expectation that a straight line can be extended to provide means of extrapolation. Actual experience indicates that these assumptions are correct provided the mentioned limits are observed. In general, it can be stated that materials with higher modulus of elasticity and higher heat deflection temperature will generate data that will lead to a line with a slope of lower degree, meaning smaller elongation per inch.

The following examples will suggest the utilization of data that will lead to successful use of plastics.

STRUCTURAL MEMBERS

The problem: A product that in principle of loading is comparable to a beam rigidly supported at both ends, and a load uniformly distributed over the supporting area with 64 psi. The use conditions require a tough material that will successfully operate at 160°F (71°C) for 5000 hours. Good flammability rating and UV stability are desired. Exposure will be to normal air. The deflection of the beam should not exceed 0.050″; the width of the beam is 0.375″ and its length is 4″. The above use requirements lead to a black, general purpose polycarbonate.

Figure 7−3 shows a schematic presentation of the problem. To solve the problem we need to determine the stress level in the beam, then look for a creep data line suitable for our conditions and establish the appropriate cross-section that will limit the ultimate deflection.

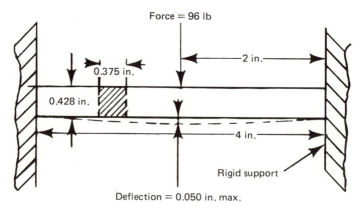

Fig. 7−3. Beam, rigidly supported at ends.

The stress (negative) in the middle is S. From the technical handbook we find

$$S = \frac{Wl}{24Z}$$

in which

W = load area × psi = 0.375 × 4 × 64 = 96 pounds.

l = length of the beam = 4″

Z = section modulus, which as a starting point shall be rectangular in shape with the depth $d = \dfrac{bd^2}{6}$

S = allowable working stress.

From suppliers' data, the change of tensile strength with temperature shows a tensile strength @ 160°F (71°C) of 7000 psi. Using a safety factor of 5 we obtain safe allowable working stress of 1400 psi.
Thus,

$$S = 1400 = \frac{Wl}{24Z} = \frac{96 \times 4}{24Z}$$

$$Z = \frac{96 \times 4}{1400 \times 24} = 0.0114$$

$$0.0114 = \frac{bd^2}{6} = \frac{0.375 \times d^2}{6}$$

$$d^2 = \frac{0.0114 \times 6}{0.375} = 0.183$$

$$d = 0.428$$

The next step is to calculate the maximum deflection in the center of the beam:

$$\text{deflection} = \frac{Wl^3}{384EI}$$

in which $W = 96$ pounds and $l = 4''$; E is the modulus at 5000 hours. The creep data show a line of 1500 psi stress at 160°F (71°C); this stress is slightly higher than the allowable working stress but is close enough to warrant the application of the appropriate modulus which is 155,000 at 160°F (71°C). We can now substitute the numbers into the deflection formula:

$$\text{deflection} = \frac{96 \times 4^3}{384 \times 155,000 \times 0.00244} = 0.042''$$

Since the moment of inertia $I = Z \times y = 0.0114 \times 0.214$ ($y = d/2 = 0.428''/2$), $I = 0.00244$.

The beam depth of 0.428" is too thick for polycarbonate moldability, therefore a "T" section or channel should be considered. Using the proportions outlined in Chapter 6, the average thickness of the rib should be 0.083" if the thickness of the base is 0.125", which is the favorable molding and property thickness of polycarbonate.

The cross-sections would be per Figs. 7-4 and 7-5.

Before making the calculations for the equivalent section modulus and moment of inertia, we should check the strain at 5000 hours and see what the equivalent deflection would be on the basis of strain in bending:

$$\text{modulus} = 155,000 = \frac{\text{stress}}{\text{strain}} = \frac{1400}{\text{strain}}$$

$$\text{strain} = \frac{1400}{155,000} = 0.009$$

The relation between strain and deflection in bending is:

$$\text{strain} = \frac{6 \times \text{deflection} \times \text{thickness of beam } (d)}{(\text{length})^2}$$

(d is obtained from the flexural test data)

$$0.009 = \frac{6 \times \text{deflection} \times 0.428}{4^2}$$

$$\text{deflection} = \frac{0.009 \times 4^2}{6 \times 0.428} = 0.056''$$

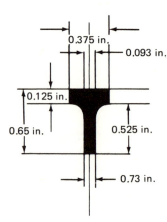

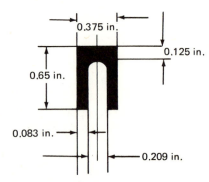

Fig. 7-4. Equivalent "T" section of beam shown in Fig. 7-3.

Fig. 7-5. Equivalent channel section of beam shown in Fig. 7-3.

This shows that deflection under constant and continuous load after 5000 hours is about 33% greater than the deflection calculated with the short-term load conditions. In order to meet the limit of 0.05" deflection, the moment of inertia of the modified section should be at least 10% higher.

We can now proceed with calculations for moments of inertia of the "T" and channel sections:

$$I_T = \frac{1}{12} \left[4bs^3 + h^3(3t + T) \right] - A(d - y - s)^2$$

$$A = bs + \frac{h(T + t)}{2} = 0.0905$$

$$y = d - [3bs^2 + 3ht(d + s) + h(T - t)(h + 3s)]/6A$$

and

$$I_C = \frac{2sb^3 + ht^3}{3} - A(b - y)^2$$

$$A = bd - h(b - t)$$

$$y = b - \frac{2b^2s + ht^2}{2bd - 2h(b - t)}$$

Substituting the values according to the figures, we have,

$I_T = 0.00328$, or 34% higher than short-term value; therefore, it has an adequate safety margin

$I_C = 0.0059$

The "T" cross-section is more than adequate and provides a considerable savings in material.

GEARING WITH PLASTICS

The subject of transmitting motion and power by means of gears, their construction, and detail requirements are fully covered in textbooks, technical handbooks, and industrial literature of gear suppliers. In this chapter we shall confine ourselves to the differences between plastic and metal gears, and shall point out the factors that must be considered for a successful application of a plastic gear.

The knowledge of gear fundamentals is a prerequisite for the understanding of where and how to insert the appropriate plastic information into the gear formulas so that the application results in favorable operation.

When discussing plastic gears, one has to recognize that we are dealing with two basic materials: The first type is a gear hobbed (cut) in a conventional manner (the same as steel gears) and made of laminated layers of fabric that are bonded with thermosetting resins under heat and pressure. The material can be in sheet form from which blanks are cut, or molded into suitable blanks ready for gear cutting. This type of gear has been in use for well over 50 years, for drives in steel mills, heavy duty machinery, or wherever the combination of laminated gears with steelmates provided the advantages of smoother operation, longer life, and lower noise levels. Timing gears for automobiles, for example, have been made on the same basic principle in very large volume over a period of many years. This type of gear has established its reliability and is considered, along with steel gears, as a proven candidate for various applications. Suppliers of laminated molded fabric materials have grades available specially suited for specific application.

The second type of plastic gear is molded, in all details, to a finished product of unfilled and filled or reinforced thermoplastic resins. These thermoplastic gears provide light weight, toughness, quiet operation, and low cost. The discussion in this chapter will deal mainly with thermoplastic molded gears.

Molded gears are used in instruments, meters, registers, windshield wiper mechanisms, and other applications where the loads are of limited magnitude and the tooth thickness through the pitch line does not exceed $\frac{3}{16}''$ to $\frac{1}{4}''$. Once the tooling is made to produce quality gears, they can be manufactured at economical costs, usually automatically, and their advantages can be further enhanced by combining with the gear such features as cams, lugs, ribs, shaft holes, webs, etc.

Careful consideration of plastic material characteristics and its processing requirements can lead to gears with the enumerated benefits and, above all, to a more successful product performance.

Formulas quoted in the following paragraphs can be found in *Machinery's Handbook*[1] under gearing, unless otherwise indicated in the text.

Considerable attention will be given to the moldability of gears. One can make all the perfect calculations and insert the necessary values for plastic gears, but if molding conditions and molding materials are not compensated for to obtain a high-quality gear, one may end up with mediocre or even unsatisfactory results. We will be discussing involute 20° spur gear teeth, but the same principles apply to other gears.

One of the first questions the designer faces in connection with a gear drive is: What is the pitch diameter of the pinion, and how many teeth is it to have? (See Fig. 7−6 for tooth nomenclature.)

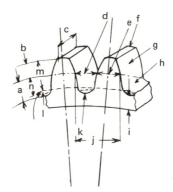

Fig. 7−6. Gear nomenclature: a, whole depth dedendum; b, addendum; c, face width; d, space width; e, tooth thickness at pitch diameter; f, top land; g, face; h, flank; i, thickness of rim; j, circular pitch; k, radius for elimination of stress concentration; l, clearance; m, working depth; n, standard dedendum.

[1]*Machinery's Handbook*, 21st ed., Industrial Press, New York, 1979.

The number of teeth in a pinion that will not have the undercut required for a working clearance of the engaging teeth of the driven gear can be determined by the formula

$$N = 2a \cos^2\phi$$

where N = number of teeth; a = 1 for American Standard full depth teeth; ϕ = pressure angle. With a = 1 and ϕ = 20°, we obtain N = 18 teeth.

Avoiding an undercut in plastic gears is important because a low-strength material cannot tolerate a decrease in strength in an area where maximum stress is produced. Furthermore, when a gear is molded, the contour in the mold that creates the undercut in the gear will interfere with the shrinkage of the tooth during solidification and thus cause a stress in the same area that is exposed to the load stress. This means a reduction in load carrying capacity (Fig. 7–7).

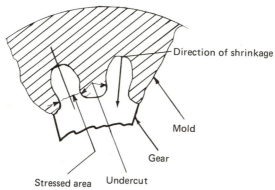

Fig. 7–7. Gear with undercut and consequent molding stress.

There are means of overcoming the undercutting of teeth in pinions with smaller number of teeth, and these methods are described in the handbooks and are equally applicable to plastic gears with the proviso that all the requirements of the plastic material are taken into account properly.

The next determination that has to be made is for the pitch diameter of the pinion and all the needed strength requirements.

Usually, the following are known: the horsepower to be transmitted, the rpm of the pinion, the center distance between driving and driven gear, and the ratio of speeds. If the center distance is not known, a reasonable allowable velocity at the pitch diameters is chosen.

Most of the gear strength calculations are based on the Lewis formulas. It has been found, however, that the results from such calculations are lower in ability to transmit power than actual experience indicates, particularly when consideration is given to the precision of gear manufacture for use at higher velocities. Since plastic molded gears can be made to close tolerance dimensions, we shall use the formula for diametral pitch determination in which the factor for the pitch-line load is reduced, depending on the accuracy with which a gear is manufactured; thus

$$P = \sqrt{\frac{3.1416 \times S_s VYK}{(HP) \times G}}$$

where

HP = The horsepower to be transmitted.

P = Diametral pitch (number of teeth/inch of diameter of the pitch circle).

S_s = Allowable static stress; in the case of plastic materials presently used for gears it is equal to about 4000 psi or about the value at the proportional limit.

V = Velocity in feet per minute as determined from available data, or if center distances are not given, recommended value is chosen depending on grade of material.

Y = Lewis strength factor depending on number of teeth and given in tables. For the 18-tooth-pinion 20° involute, the value is 0.308.

K = Factor 3−5 for the face width. Such a width is established by multiplying this factor with the circular pitch. In our example, we shall choose a $K = 3$.

G = Velocity factor = $427(78 + \sqrt{V})$ for accurate gears that will work at pitch-line velocities up to and above 2000 ft/min.

The solution of the diametral pitch formula requires the selection of a permissible velocity for those cases where the center distance of the gears is not given. As a guide for velocity selection, we can use the recommendation of the American Gear Manufacturers Association (AGMA) for the nonmetallic gears (laminated fabric bonded and molded with phenolic resin), namely, that the velocities should be 600 ft/min or higher. The action of the nonmetallic gear is similar to that of the thermoplastic gear, therefore, the AGMA values can be safetly applied. It is also their recommendation that the diametral pitch of the nonmetallic gears bears a relationship to horsepower, pitch line velocity, and torque. These values are given in the technical handbooks.

With thermoplastic gears, in addition to the above guidelines, there is another important consideration, namely, that of tooth thickness at the pitch line.

From a processing standpoint, thermoplastics have a favorable maximum thickness in which they can be reproduced in an economical and trouble-free manner that will generate high strength parts. That thickness usually is on the order of 0.150" for unfilled resins and 50% greater for filled resins. In these thicknesses for gear teeth, they retain a certain flexibility that permits deflection of teeth and makes possible the engagement of more than one tooth at a time. This action distributes the load over greater tooth areas. In metal gears a comparable deflection would most likely exceed the elastic limit, and cause tooth deformation, associated problems of excessive wear, and short life. Too large a flexibility in teeth can cause an excessive amount of teeth engagement with the side effect of binding. A tooth thickness one-third that of the favorable plastic thickness, or 0.050" through the pitch line, may be the beginning of too much flexibility for some thermoplastics.

Since the velocity factor G was predicated on a precision-molded plastic gear, we should examine the design and molding requirement that will lead to such a gear. The following elements will be the determining factors for a high-quality plastic gear.

1. *Tooth thickness.* The tooth thickness at the pitch line should be approximately 0.150". This applies to materials presently most frequently used for gears and the polymers that are not filled nor reinforced. If tooth thickness is higher than the 0.150" requirement, then the base on which the tooth is supported is usually heavier, and the net result is a tendency for porosity and voids. These drawbacks can be minimized with higher mold temperatures and longer cycles, but they will adversely affect cost and partially detract from one of the advantages of a molded gear.

When the polymers are filled or reinforced, the tooth thickness can be increased by 50% up to 100%, depending on the percentage of filler. With 10% filler as the low end and 40% as the high limit, the acceptable thickness can be proportionately adjusted. The tendency for void formation of filled material is minimal, but the fillers being abrasive in comparison with the matrix material usually call for an additive that will counteract such undesirable wear action. Filled materials, on the other hand, provide higher physical properties, therefore being another item for consideration in evaluating materials for gear selection.

2. *Stress concentration factor.* Notch sensitivity of plastic materials makes it necessary to provide a radius between the teeth, which starts as a tangent at the working depth of the teeth and continues to or even slightly beyond the clearance. (See Fig. 7–6.) This arrangement eliminates a notch-sensitive condition and prevents a stress concentration in a critical area.

3. *Feeding the gear cavity.* With crystalline materials, the flow pattern of the material in the cavity must be such that it starts in the center of the gear and radiates uniformly to all the teeth. This creates a uniform shrinkage of the complete gear, and we obtain a gear with greatest accuracy in tooth shape and spacing; this means that the gating should be of the disc or diaphragm type. (See Fig. 7–8.) Feeding the cavity eccentrically, even with three gates uniformly spaced, will produce low points that will change the pitch circle from round to, say, three arcs connected with flattened portions of different arcs. In the case of amorphous materials with low shrinkage values, the center gating is not quite as significant as it is with the semicrystalline or crystalline types.

The feeding of filled or reinforced polymers must be done from the center to the outside diameter whether the base material is crystalline or amorphous.

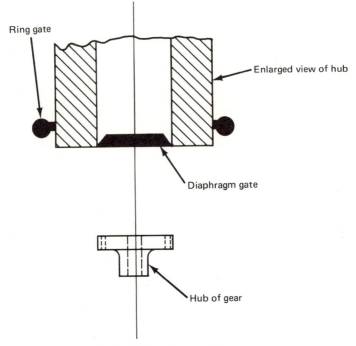

Fig.7−8. Gating for precision gear.

Eccentric gates will cause the reinforcing material to arrange itself in a directional manner in some areas, while in others it will retain its random mixture. This condition will cause differences in shrinkage and bring about a gear that is out-of-round. Furthermore, a gate away from the center will cause weld lines which, in turn, can reduce the strength of material by about 20%.

4. *Face width of gear.* When a thick (deep) plastic part is molded, there is normally a need for some taper that will allow the part to be ejected from the cavity in a stress-free condition. On the gear, the face width would be involved in the taper requirement. A precision gear must have a straight face in order to ensure that a full load is carried by the calculated value of face width. To accomplish this, the mold that produces the gear must have a very low surface finish reading (3−5 microinches) with the polishing lines following the direction of ejection of gear from mold. Furthermore, the plastic material should have a light lubricant as an additive that tends to migrate to the surface of the gear. The slight shrinkage and high directional finish of the cavity permit the stress-free removal of the gear. Keeping the above in mind, it would be advisable to have the face width on a low side of a normal practice.

5. *Webbed gears.* The proportions of gear tooth thickness to web and rim thicknesses should be such as to avoid the danger of porosity and void formation. Under the heading of design features, the rib thickness in relation to its

base is shown as two-thirds of base thickness. If we follow the same basic approach and consider the tooth thickness t as the rib, we should have a rim and web thickness of $1\frac{1}{2}t$. (See Fig. 7–9.) The general principle of uniform wall thickness throughout the part should be adhered to in the interest of keeping molding stresses to a minimum.

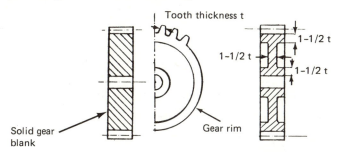

Fig. 7–9. Webbed gear vs solid blank gear.

Having explained all the factors in the diametral pitch formula, we can now proceed with the calculations of our example and determine tooth thickness. Substituting appropriate values in the formula, we have

$$G = 427 \, (78 + \sqrt{600}) = 43{,}763.23$$

$$P = \sqrt{\frac{3.1416 \times 4000 \times 600 \times 0.308 \times 3}{0.25 \times 43{,}763.23}} = 25$$

The nearest standard diametral pitch is 24 and the tooth thickness for it is

$$\frac{1.5708}{P} = \frac{1.5708}{24} = 0.065$$

Having selected an allowable velocity of 600 ft/min, we can also use this factor as a basis for determining tooth thickness:

$$V = 0.262 \times D \times R$$

where D is the pitch diameter and R is the revolutions per minute. In our example, the values are

$$600 = 0.262D \times 1800$$

$$D = \frac{600}{0.262 \times 1800} = 1.272''$$

and since

$$D = \frac{N_p}{P} = \frac{18}{P} = 1.272$$

where N_P is the number of teeth in the pinion

$$P = \frac{18}{1.272} = 14 \text{ diametral pitch}$$

The tooth thickness for this = 1.5708/14 = 0.112″. The 0.065″ thickness may be too flexible and tends to bind especially if operating temperature would be above 73°F (23°C).

For this reason it would be prudent to use the average of the two thicknesses or (0.065 + 0.112)2 = 0.088 that would correspond to 0.088 = 1.5708/*P* or *P* = 1.5708/0.088 = 17.85, or an 18 diametral pitch with a tooth thickness = 1.5708/18 = 0.087″, which is being adopted for the example and is very practical from a quality point of view as related to processing and gear performance.

The final consideration for tooth thickness is the problem of operating clearance known as "backlash."

Backlash and Working Clearance

Backlash is defined as the measurement by which the space between teeth exceeds the thickness of the engaging tooth on the pitch circle. Backlash is necessary to prevent simultaneous contact on the two sides of the space and thus eliminates the possibility of binding. Backlash, tip relief, and similar arrangements are the means of providing satisfactory working clearance and thus minimizing excessive wear and noise.

The factors that cause a gear to mesh tightly are: (1) tolerance on concentricity of shaft hole with pitch diameter, (2) tolerance on center distance, (3) AGMA tolerance on quality, (4) the coefficient of thermal expansion, and (5) change in dimensions due to moisture absorption, which is a consideration in some materials. The first three apply to gears of any material. Item 4, and in some cases Item 5, for plastic gears deserves special consideration since most plastics expand about 10 times as much, for example, as steel.

By way of specifics in which these factors are applied, we will assume that the 18*P* pinion established early in this chapter will have 1800 rpm and a speed reduction of 3 to 1.

1. Maximum tolerance on concentricity of hole with OD of gear	= 0.0005
2. Tolerance on center distance	= 0.0025
3. AGMA quality (total composite error of Class 3 gears)	= 0.00025
4. Thermal expansion: 73°F (23°C) [operating at 150°F (66°C)]	= 0.0041
Cumulative variables	0.00735

This difference in operating center distance from theoretical center distance will force the gears into closer mesh:

$$\text{Theoretical center } C = \frac{n + N}{2P} = \frac{18 + (3 \times 18)}{2 \times 18} = 2$$

where C_1 is the center distance at which the expanded teeth and tolerances will cause closer meshing of gears.

$$C_1 = C - \text{Cumulative variables} = 2 - 0.00735 = 1.99265$$

Since, according to the handbooks,

$$C_1 = \frac{\cos \phi}{\cos \phi_1} C$$

in which ϕ is the involute function of the standard pressure angle, in this case 20°, and ϕ_1 is the involute function of operating pressure angle when gears are meshing tightly:

$$\cos\phi_1 = \frac{\cos\phi \times C}{C_1} = \frac{0.93969 \times 2}{1.99265} = 0.943156$$

$$\phi_1 = 19\frac{26}{60} = 19.433$$

We can now figure what the combined thickness of driving and driven teeth should be in order to have enough backlash to operate freely. From the handbooks,

$$\text{inv. } \phi_1 = \text{inv. } \phi + \frac{P(t + T) - \pi}{n + N}$$

in which $t + T$ is the combined thickness of teeth, $n = 18$, $N = 54$

$$\text{inv. } \phi_1 = 0.013634$$

$$= 0.014904 + \frac{18(t + T) - \pi}{72}$$

$$72 \times 0.013634 = 0.014904 \times 72 + 18(t + T) - \pi$$

$$\frac{72(0.013634 - 0.014904) + \pi}{18} = t + T = 0.1693$$

$$\text{Standard tooth thickness} = \frac{1.5708}{18} = 0.087$$

$$T_p + T_g = 0.174$$

The difference is 0.0047 or 0.005″ backlash.

$$T_p = 0.087 - 0.002 = 0.085 \text{ and } T_g = 0.087 - 0.003 = 0.084.$$

When cavities for gears are produced by the electric-discharge machining method, the backlash as well as tip clearance or other tooth modification can readily be attained without the need for shaving or similar operations of tooth form.

We have up to this point established what it takes to make a quality plastic gear and now we can proceed with the strength determinations.

Strength Factor 1. Contact ratio is defined as intermeshing gears in which one pair of teeth remains in engagement until the following pair is in position to carry the load. Plastic gears, being made of a flexible material in comparison with steel, should normally have a relatively higher contact ratio than would be the case with metal gears. Too high a ratio may cause binding which would call for corrective measures such as increasing the center distances or substituting a stiffer material. The formula for contact ratio is:

$$CR = \frac{\sqrt{R_o^2 - R_b^2} + \sqrt{r_o^2 - r_b^2} - C \sin \phi}{p \times \cos \phi}$$

where

$R_o = \frac{1}{2} \times \dfrac{N_p + 2}{P}$, outside radius of first gear

N_p = number of teeth and P is diametral pitch.

$r_o = \frac{1}{2} \times \dfrac{N_g + 2}{P}$, outside radius of second gear

$R_b = \frac{1}{2} D \times \cos \phi$ = is the base circle radius of the first gear, where D is pitch diameter

p = circular pitch and ϕ is the pressure angle of 20° involute gears and C is the center distance of gears.

Substituting the values for our example, we find a contact ratio of 1.7, well above the considered minimum of 1.4. The factors appearing in the formula for contact ratio have no affect on material properties, therefore, they require no special consideration.

Strength Factor 2. Tooth strength (based on Lewis formula). The power transmitting capacity is

$$HP = \frac{LV}{38,000}$$

where HP = horsepower to be transmitted

L = maximum safe tangential load at pitch diameter

V = velocity in ft/min at the pitch diamter = $0.262 \, DR$

D = pitch diameter in inches and R = revolutions per minute

L = SFY/P, where S is allowable unit stress for material at given velocity

F = width of face in inches = K-factor × circular pitch (In our
example, K-factor = 3 and circular pitch = 3.1416/P =
3.1416/18 = 0.1745)

Y = outline factor (Lewis factor)

From the handbooks, P is the diametral pitch;

$$S = \frac{Ss \times 600}{600 + V}$$

where Ss = static stress or 4000 psi for the thermoplastics used for gears. The
determination of this stress is made by applying the value of the proportional
limit from the stress–strain data; thus

$$L = \frac{SFY}{P} \quad \text{or} \quad L = \frac{Ss \times 600}{600 + V} \times \frac{F \times Y}{P}$$

This formula points out that the tangential load is influenced: (1) by a
velocity factor that reduces the static working stress; (2) by the Lewis outline
factor whose value increases with the number of teeth; (3) by the diametral
pitch, i.e., the higher the number, the lower the tangential load. The only factor
that increases the tangential load capacity is the width of the face.

In the example that we selected, the horsepower capacity would be

$$S = \frac{4000 \times 600}{600 + 471.6} = \frac{2,400,000}{1071.6}$$

$$= 2239.64 \text{ psi}$$

$$V = 0.262 \times 1 \times 1800 = 471.6$$

$$L = \frac{2239.64 \times 3 \times 0.1745 \times 0.308}{18} = 20.06 \text{ lb}$$

and

$$HP = \frac{20.06 \times 471.6}{33,000} = 0.286$$

The selected and calculated values indicate adequate horsepower capacity
and a safe tangential load in line with AGMA recommendations.

Strength Factor 3. Dynamic or maximum momentary load. For this type of
a load the Buckingham dynamic load formula is being used:

$$L_d = \frac{0.05V(FB + Lt)}{0.05V + \sqrt{FB + L_t}} + L_t$$

where

$$L_t = \frac{HP \times 33{,}000}{V}$$

Except for the factor B, the deformation factor, the remaining factors are as before.

The major factor that influences the dynamic load is the deformation factor B. The forces that contribute to the deformation factor are due to pitch and profile error, the inertia of the rotating mass and possibly misalignment in mounting. In precision plastic gears these contributors that increase the momentary load are much smaller than is the case with steel gears. Furthermore, plastic materials have a much higher degree of tolerance for deformation than steel; therefore, the factors entering into the deformation formula will not have the degrading effect found in steel. If, applying the tables from the handbooks, we obtain high values for momentary loads in comparison with the static load, we should consider this an expected occurrence in view of the difference in deformation characteristics between plastics and steel.

The difference in deformation can be best appreciated by comparing the elongation of steel at its proportional limit of 0.001 in./in. with that of polycarbonate at its proportional limit of 0.0116 in./in. or a multiple of 12 higher.

The handbook table shows the deformation factor of steel against steel, for error in action varying conditions. For precision plastic gears, the error in action can be 0.0007″, which would correspond to a deformation factor in steel of 1162. The deformation should be in proportion of the modulus of elasticity of the different materials.

$$\frac{E_p}{E_{st}} \times 1162 = \frac{3.45 \times 10^5}{30 \times 10^6} \times 1162 = 13.35$$

The plastic deformation factor is 13.35. Therefore, the momentary load in our example is

$$L_d = \frac{0.05 \times 471.6 \, (3 \times 0.1745 \times 13.35 + 17.49)}{0.05 \times 471.6 + \sqrt{3 \times 0.1745 \times 13.35 + 17.49}} + 17.49$$

$$= 36.38 \text{ pounds}$$

Using the Lewis formula for reducing the allowed stress by applying the reciprocal value to the dynamic load, we have

$$L_d = L_t \times \frac{600 + V}{600} = 17.49 \times \frac{600 + 471.6}{600}$$

$$= 31.23 \text{ pounds}$$

Since we are calculating a momentary load to see if it is below or equal to the static load, both methods should be employed.

Strength Factor 4. Static beam strength of teeth. The HP = $(L_b \times V)/G$ and $L_b = SsFY/P$. The letter designations are the same as previously indicated.

The static beam strength is used as a basis for comparison with other calculated values under operating conditions, which obviously should be lower than those at standstill. In our example,

$$\mathrm{HP} = \frac{L_b \times V}{G} = \frac{35.83 \times 471.6}{43,763.43} = 0.386$$

and

$$L_b = \frac{4000 \times 3 \times 0.1745 \times 0.308}{18} = 35.83 \text{ pounds}$$

Strength Factor 5. Wear load limit. The Buckingham formula for wear is

$$L_w = D_p F K Q$$

where L_w is the limiting static load for wear in pounds; D_p pitch diameter for pinion; K is the load stress factor,

$$K = \frac{S_c^2 \times \sin A}{1.4}\left(\frac{1}{E_1} + \frac{1}{E_2}\right)$$

F is the face width; Q is the ratio factor = $2N_g/(n_g + N_p)$; N_g is the number of teeth in gear; and N_p is the number of teeth in pinion. This formula applies to external spur gears. The new factors in this wear formula are the stress factor that depends on surface finish of the teeth and how well the materials run against each other, i.e., the relative coefficient of friction or, in the case of metals, relative hardness. The stress factor equation expresses the stress factor in terms of stress and strain conditions, and these have proven themselves in practice over periods of many years. The ratio factor indicates that the more teeth involved, the higher the wear load limit.

Essentially, the wear load limit is a function of the capacity for endurance limits by the material in the teeth. For this reason it is one of two limiting loads in gear strength determination. The second is the tooth strength (Lewis formula) described in Point 2, and whichever is lower is the one to be applied as the limiting number.

In our example, the conditions will be:

$$L_w = D_p F K Q$$

where

$$K = \frac{S_c^2 \times \sin A}{1.4}\left(\frac{1}{E_1} + \frac{1}{E_2}\right)$$

where $S_c = S_s = 4000$, A = pressure angle of $20°$, and $E_1 = E_2 = 345,000$ (average) psi;

$$K = \frac{4000^2 \times 0.342}{1.4} \times \frac{2}{345,000} = 22.66$$

$$Q = \frac{2.54}{54 + 18} = 1.5$$

$$L_w = 1 \times 3 \times 0.1745 \times 22.66 \times 1.5 = 17.79$$

The required load from the Lewis strength formula is 20.06, therefore, in order to increase the L_w to that value, we shall increase the face width from 3×0.1745 to 3.5×0.1745, which will change the L_w to $17.79 \times 3.5/3 = 20.75$.

There are additional elements that contribute to gear life. Some of these are as follows:

1. The operating temperature of the gears should be allowed for, using the strength data of the material at the temperature in question whenever a formula calls for a stress value.

2. Plastic gears should be enclosed for protection against atmospheric dust and dirt to eliminate their abrasive effect.

3. Lubrication should be provided whenever feasible, otherwise a derating factor should be included. Suitable lubrication may increase gear life by 50%.

4. Material fatigue at a required stress level and for a specific diametral pitch should be established.

5. The nature of load, whether intermittent or continuous, whether a shock load or steady smooth load, will call for appropriate derating factors.

6. The coefficient of friction factor between mating materials will also influence life of the gears.

In conclusion, it should be noted that unless there is experience with a prior similar application, prototyping of plastic gears and thorough testing is the appropriate step. If keyways are employed, sharp inside corners should be avoided due to high notch sensitivity of plastics. A radius of $\frac{1}{32}''$ in the corner is the minimum with a corresponding change in the key.

BEARINGS

Similarly to gears, some plastic bearings have a long history of successful performance. The laminated fabric, bonded with phenolic resin incorporating antifriction ingredients and cured under heat and pressure, has given excellent service when properly applied. This group of bearings has a low coefficient of friction, antiscoring properties, and adequate strength for use in steel mills and other heavy duty applications and is well established in the industry. The thermoplastic materials that are considered for bearing applications are for relatively light duty work such as food mixers, adding machines, and similar devices.

Some thermoplastics have inherent lubricating characteristics that can be enhanced by additives such as Teflon, molybdenum disulfide, and others. Other thermoplastics, as well as thermoset resins, by the addition of Teflon and/or molybdenum disulfide, become excellent candidates for bearing materials. Supplying bearing materials of the composite type has become a specialty that is followed by several companies.

When designing a plastic bearing, one must recognize that the primary cause for favorable performance will be keeping the frictional heat to a low value and having prevailing conditions that lead to dissipation of such generated heat. It has been established that the factors in a bearing system contributing to frictional heat are the pressure P exerted on the projected area of the bearing and the velocity V or the speed of the rotating member. This is known as the PV factor, and maintaining the values within the prescribed limits of each material will lead to a successful bearing. (Fig. 7−10.)

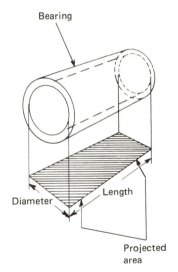

Bearing

Diameter

Length

Projected
area

Fig. 7−10. Projected area of bearings.

The elements that contribute to the limiting of the PV factor are magnitude of pressure, speed of rotation, coefficient of friction of mating materials, lubrication, clearance between bearing and shaft, surrounding temperature, and surface finish, as well as hardness of the mating materials. Bearing wall thickness is also an element in the PV factor since it determines the heat dissipation.

By treating individually each of the above elements, we find the following: The limit of the *PV* factor for each material or the internally lubricated composite for constant wear of bearing is usually available from the supplier of the plastic. Neither the pressure nor the velocity should exceed a value of 1000. Thus, if the *PV* limit of acetal is 3000, the *PV* factor could be 1000 ft/min times 3 pounds or 1000 pounds times 3 ft/min at the extreme, provided heat conditions resulted in uniform rate of wear.

The coefficient of friction data are also available from supplier.

Lubrication whether incorporated as a composite material or provided by feeding the lubricant to the bearing will raise the *PV* limit 2.5 or more times over the dry system. The possibility of a rusting shaft should be guarded against in order to prevent excessive bearing wear.

Clearance between shaft and bearing will be large in comparison with, for example, a bronze bearing, mainly due to the high coefficient of expansion of plastics—roughly 10 times that of steel. The average allowed clearance is 0.004–0.005″ per inch of bearing diameter. This is needed in order to counteract the tendency to "close" in the bearing ID when temperatures rise or assembly conditions cause a decrease in the shaft hole. The surface roughness of the shaft can play a large part in heat generation in the bearing. The protruding ridges from machining on the shaft act as minute cutting tools and disturb the smooth surface of the bearing thereby creating heat due to interference with the displaced material. A surface reading of 5 microinches is almost mandatory on the bearing portion of the shaft in order to ensure good plastic life. Surface hardness of at least 300 Brinell on the shaft is another favorable feature for smooth operation. It prevents the pick-up of loose particles that can abrade the polymer material.

Wall thickness of the bearing should be on the low side to facilitate heat transfer into the housing, and yet it should be of sufficient value to facilitate effective manufacturing that will be conducive to a quality product. The thickness should be in proportion to shaft diameter, starting with 0.05″ on the low side and ending with 0.2″ on the high side. For filled materials, these thicknesses can be 50% higher.

Bearing length should be equal to or less than 1.5 diameters.

Concluding the remarks on plastic bearings, one must appreciate the numerous variables that enter into their design, and for that reason a prototype test is very much in order.

8. Joining or Assembly Techniques

The joining of plastic components to each other or to components of other materials is a frequent occurrence in product assembly. The success of a specific technique will depend on whether, as a by-product of the technique, sizable stress levels in the plastic part result. Guarding against potential stresses in the assembly of parts is a very important aspect of complete product design.

SCREWS

A self-threading screw into a plastic material is the most desirable type since its action can be compared to a tapped hole. For best holding power in thermoset as well as thermoplastics, the screw should be of the coarse pitch design (similar to a sheet metal screw) and at the start of the threads should have a slot designed to act as thread cutting edge. The holes that receive the screw should be of a size recommended by the screw manufacturer for a specific plastic material. An accurate hole size determines the optimum holding power. A drill diameter corresponding to a hole size will not necessarily produce the desired hole size in the plastic. After drilling, shrinkage can be appreciable in some materials; therefore, a molded hole should be specified wherever feasible. The torque required for each set of conditions has to be established experimentally to avoid damage to the created threaded assembly. Another type of screw that can provide good holding power in plastics is the type with a deep and thin thread, which is designed to displace the engaging material rather than cut it. If assembly conditions are such as to cause the plastic material to soften and flow readily, then the result should be satisfactory, otherwise an undesirable stress condition is created that will cause failure. Heating the screw, for example, to about 300°F (149°C) while driving it into position and providing a space for the displaced plastic can lead to favorable results. New screw designs are brought on the market with claims for advantageous engagement into plastics. The question in each case is: Will the assembly result in a low stress level of the plastic material and how much holding power is obtained?

HOLDING WITH FORMED HEADS

A holding head is similar to the head formed during riveting except that in the plastic part there is a protruding stud that fits through a hole in the part to be joined and the head is shaped over it. It is an economical method of joining. The head can be shaped by spinning or ultrasonic forming. (See Fig. 8–1.)

The spinning operation consists of high-speed rotating and suitably shaped tool that creates frictional heat which will permit the stud to conform to the configuration in the tool. Pressure exerted on the tool and the time of rotation

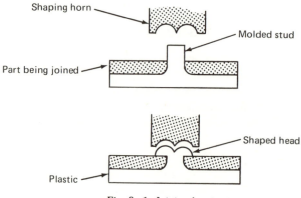

Fig. 8−1. Joining by riveting.

are accurately controlled. The spinning device can produce joints, at high speed, of good quality.

Ultrasonic forming and welding is the fastest growing assembly technique. It is a very rapid operation of about 2 seconds or less and lends itself to full automation. In this process high-frequency vibrations and pressure are applied to the parts to be joined, heat is generated at the plastic causing it to flow, and, when the vibrations cease, the melt solidifies. The heart of the ultrasonic system is the horn, which is made of a metal that can be carefully tuned to the frequency of the system. The manufacture of the horn and its shape is still a state-of-art process and is normally developed by the manufacturer of the equipment. A schematic outline of the equipment is shown in Fig. 8−2.

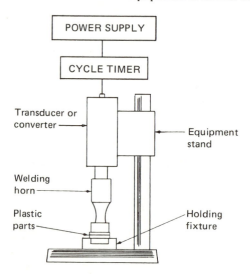

Fig. 8−2. Schematic outline of an ultrasonic welding machine.

The results from this operation are not only economical, but also most satisfactory from a quality standpoint.

WELDING

Frictional spinning can be applied to welding of two plastic parts with circular joints. It is especially suitable for large parts where ultrasonic welding may be impractical or equipment cost prohibitive. In this operation the faces to be joined are pressed together while one part is rotated and the other is held in a fixed position. Frictional heat produces a molten zone that becomes a weld when rotation stops. When alignment is precise and centering means are incorporated in the parts, the result is a good joint in terms of strength and appearance. The approximate parameters are 40 to 50 feet per minute peripheral speed and 300 to 400 psi pressure.

Ultrasonic welding's principle of operation is the same as described previously, except that the design of the joining surfaces requires special attention. The important feature in ultrasonic welding is the energy director that consists of an initial small contact area through which the flow of energy is started. Figure 8−3 shows a butt joint between joining surfaces; Fig. 8−4 displays a more positive method of joining where the self-aligning feature is included; Fig. 8−5 shows the "shear" joint system, which has proven to be very effective for producing leak-proof joints. Variations of this design can be adopted readily to larger parts without the need of resorting to proportionately larger welding facilities.

The cycle times are fast—less than 2 seconds—and energy consumption is low due to the fact that only a thin layer of material on both components is softened, from which a good welding joint is obtained.

Welding of plastics can also be accomplished with the aid of a torch or heating air gun. These methods are usually employed for experimental purposes.

Energy director

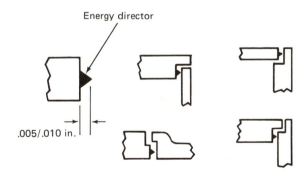

.005/.010 in.

Fig. 8−3. Ultrasonic welding of butt joints.

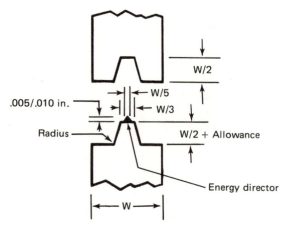

Fig. 8−4. Ultrasonic welding of self-aligning joint.

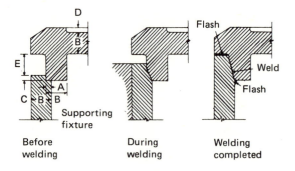

Dimension A: 0.016 in. (0.4 mm). This dimension is
constant in all cases.

Dimension B: This is the general wall thickness.

Dimension C: 0.016–0.024 in. (0.4–0.6 mm). This recess
is to ensure precise location of the lid.

Dimension D: This recess is optional and is generally
recommended for ensuring good contact
with the welding horn.

Dimension E: Depth of weld = 1.25 to 1.5 B for maximum
joint strength.

Fig. 8−5. Ultrasonic-welding "shear" joint design.

SUMMARY

The joining of plastics is only limited by the ingenuity and skill of the designer. The only precaution that has to be exerted, whether it be cementing or snap fitting or any method of forming, is that no stresses are generated during the operation.

In thermoplastics the preferred method is to have the material softened so that it can flow and adjust to the new condition. The softening need be only a few mils thick to have a favorable result. The plastic does not "care" where the heat comes from as long as the temperature is within its melt limits.

9. Description of Processing Plastics

Most of the engineering products are currently designed to fit processes that have proven themselves as reliable means to consistently produce parts of desired quality. The processes described in this chapter are intended to convey the underlying principles of some of the most frequently used methods for engineering products. The detailed advantages of any of the methods described here or others can be obtained from equipment manufacturers or literature on the type of process of interest. When a product design leads to one specific method, it is suggested that one familiarize himself with it in order to ensure that the chosen method is most conducive to favorable product results. A further suggestion is that, before the design details are finalized, the contemplated process be firmed up, so that any requirements inherent to the process be incorporated in the finished design of the product.

Injection molding. Injection molding is by far the process that is most widely employed by designers. The main reason for this preference is that dimensions and shapes can be accurately controlled and predicted. Figure 9−1 shows a schematic outlining of the injection machine where the material path from hopper to mold is shown and where the part is formed. The process can be described as follows: Plastic raw material is loaded into a hopper. From there it drops by gravity onto the feeding portion of the reciprocating extruder screw. The flights of the rotating screw cause the material to move through a heated extruder barrel (heating cylinder) where the plastic material is softened and made fluid, so that it can be fed into the shot chamber. During the time that material is moved forward, the nozzle that contacts the mold is blocked by the preceding shot into the mold. As the screw rotates and moves the material to the front of the shot chamber, a pressure is generated by the

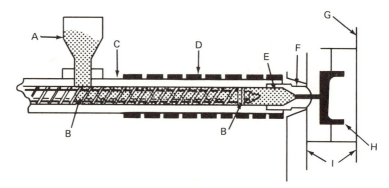

Fig. 9−1. Schematic outline of injection molding. A, material hopper; B, check rings (valves); C, heating cylinder; D, heating units; E, shot chamber; F, nozzle; G, clamp; H, molded part; I, platens.

flowing plastic which causes the screw to retract and thereby create space for the incoming material. The screw continues rotating and delivering material until the required volume for the shot is completed. At this point a properly set limit switch stops the feeding and retracting action of the screw and the screw converts itself into a plunger that shoots the fluid plastic into the mold completing the injection phase of the operation. The non-return valve prevents any possible back flow of material into the chamber. Injection takes place at high pressure (up to 20,000 psi), meaning that the mold has to be tightly clamped by the press to prevent leakage of material during the shot. Following injection, the cooling (curing) solidification of the plastic takes place while, concurrently, the next shot is being prepared. When the parts attain the necessary rigidity by a predetermined time setting, the press opens the mold, causes the ejection mechanism to go into action, which removes the plastic from the mold, and another cycle of operation begins. The clamping of the mold is performed under pressures of 2−7 tons per square inch of projected molding area. The range of pressures depends on the fluidity of material during injection. All the machine actions are governed by timers and limit switches that bring about the faithful repetition of the operations.

If parameter settings for every element in the cycle are established by qualified personnel and, subsequently, are carefully reproduced and maintained, one can expect very accurate dimensions and excellent reproducibility of quality. Failure to recognize the importance of accurate parameter settings can lead to all sorts of product problems. It should be kept in mind that in processing plastics the factors of time, temperature, and pressure are the guiding signs for quality determination. *Under time factors* are included rate of injection, duration of pressure application, time of cooling, time of material plastication, and rpm of extruder screw. *Under pressure factors* are injection high pressure and low pressure, back pressure on extruder, and pressure loss before entering cavity, which can be caused by a variety of factors. *Under temperature factors* are mold temperature (cavity and core), barrel temperature, nozzle temperature, temperature due to back pressure, temperature due to screw speed, frictional heat, and temperatures causing gas formation.

This large number of variables, if not controlled properly, can cause part changes, and, for that reason, the idea of process specification for some critical part may be considered or even the requirement of including process control facilities may be in order.

The injection process is as good as the recognition and controlling of the parameters. It is the process that will produce parts to close tolerances and that will generate good appearance and good quality overall.

The injection process is also employed for thermosetting materials.

A relatively new version of injection molding is the so-called *RIM* (*reaction injection molding*) *process*. The RIM process starts with basic chemicals that make up the material, mixes them, and shoots the mixture into the mold where the chemicals react under heat and pressure and form the finished product.

The process eliminates the step of forming the plastic and preparing it for molding. Wherever applicable, the RIM process should be more economical and provide the freedom of manipulating the chemicals for the purpose of compounding grades to suit specific applications.

Compression molding. In compression molding a charge of material is placed into the mold cavity where, under heat and pressure, the material is plasticized and caused to fill out the configuration of the cavity. (See Fig. 9–2.) The charge can be in the form of loose raw material or compressed into tablets. When the charge is preplasticized (heated and softened) and dropped into the cavity, the cycle time can be very close to that of injection molding. The overall properties of a compression-molded part are usually more favorable than by other processes.

Each process has its good points and weaknesses, and only a detailed evaluation in relation to a specific part can lead to a decision as to the direction to choose. In compression molding, the outstanding benefits are a high degree of part density and a very low tendency toward warpage and distortion.

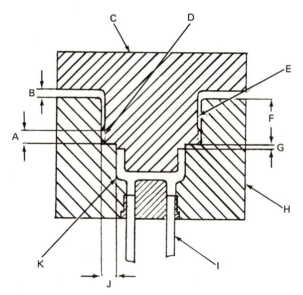

Fig. 9–2. Schematic outline of a compression mold: A, shut-off height; B, overflow space; C, force or male core; D, overflow grooves; E, clearance and space for overflow; F, loading space for raw material; G, provision for vertical flash; H, cavity; I, KO pins; J, shut-off land; K, part.

Thermoforming (Fig. 9–3). Thermoforming is a process in which a thermoplastic sheet is heated to a softening point and then drawn or forced over an open mold. The mold can be male or female, and the forming force is either vacuum and/or pressure. Broadly speaking, it can be compared to metal stamping, except for the heating of the sheet material.

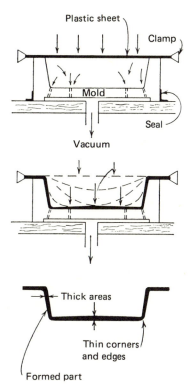

Fig. 9-3. One of the many variations of thermoforming.

After the product is formed, it is cooled to a specific temperature to prevent distortion. The last step in the operation is trimming to the outside contour of the product. The principal use for thermoformed products is packaging food products, and it is being extended to industrial applications. Whenever thickness reduction due to shape forming is tolerable, thermoforming can be evaluated and may result in its favorable consideration.

Tooling in comparision with injection molds is relatively inexpensive, although cycle times are somewhat longer.

Extrusion (Fig. 9-4). The extrusion process has many subdivisions and is used for producing plastic film sheeting, pipe, and other continuous lengths with irregular cross sections, and for wire coating and other coating systems.

The operation of the extruder is as follows: Plastic material is fed from the hopper by gravity onto the cylinder of an extruder screw. The constantly turning screw feeds the material forward through the heating cylinder, where additional heat is gained from the shearing energy of the screw until a homogeneous melt is attained. Prior to leaving the cylinder, the plastic passes a braker plate and screen pack. This unit causes back pressure buildup in the cylinder, filters out contaminants, and brings about a laminar flow of the plastic. The die

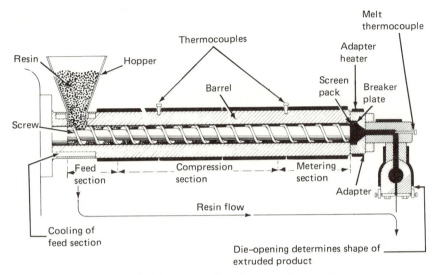

Fig. 9−4. Schematic outline of an extrusion machine.

that shapes the extruded product is located in the adapter of the extruder, and when the plastic leaves the die, it enters the sizing unit followed by a cooling unit where the temperature is reduced to the point of product stability. After the cooling unit, a take-off or hauling unit keeps the material moving up to the stage where a traveling cut-off saw sizes the material to required lengths.

Blow molding. (Fig. 9−5). In blow molding the extruder screw is applied to plasticize the material as it is moved through the heated chamber and additionally picks up the heat from shearing to produce a homogeneous melt that enters the die head directly or through an accumulator into the die head.

A tube of molten plastic called the *parison* exits from the die head. A mold whose cavities are shaped to reproduce the finished product clamps around the parison, while on the inside of the parison there is a compressed air connection that provides the force to expand the hollow tube and fill out the cavity. Upon contacting the cooled mold, the plastic becomes rigid enough to retain the cavity shape and is ready for ejection. With the mold empty, a new cycle begins.

The process is used for liquid containers and similar hollow items, where the tolerance of wall thickness is not tight.

Tooling cost is relatively low, using injection molds as a standard of comparison.

Rotomolding. Rotomolding is used for making hollow parts for a variety of applications. The process is carried out along these lines: A predetermined charge of material is placed into a mold and closed for the operation. The mold, usually constructed of lightweight material, is rotated and tumbled (the rotation is simultaneous in two directions perpendicular to each other); it is heated by

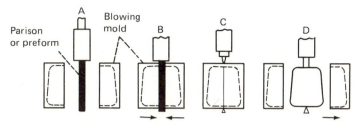

Fig. 9−5. Schematic outline of blow molding steps. The extruder used in blow molding is similar to the one in Fig. 9−4 except that the front is altered to adopt a programming head, which produces a properly sized preform for the product. A, The molten, hollow parison or preform is placed between the two halves of the blowing mold. B, The blowing mold closes around the parison. C, The parison, still molten, is pinched off and inflated by an air blast that forces its walls against the inside contours of the cooled mold. D. When the piece has cooled enough to have become solid, the blowing mold is opened and the finished piece is ejected.

forced hot air in an oven. As the walls of the mold pick up heat, the plastic begins to stick to the inside of the mold, thus forming a hollow part. After the material is fully distributed, the resin particles continue to bond to each other until a uniform layer of part thickness is formed on the inside of the mold.

The next step is for the indexing mechanism to place the mold into a cooling zone. The rotation is continued in the cooling zone where a controlled method of heat dissipation is employed to ensure desired properties and appearance in the product. At this point it is possible to introduce pressure on the inside of the part to ensure shrinkage uniformity and to prevent distortion. When the mold reaches a temperature suitable for part removal, rotation stops, and the indexing mechanism places the unit into the unloading position.

Some of the products made by this process are toys like hobby horses, mannequins, arm rests, head rests, boat bumpers, and similar hollow items.

Summary. In every one of these processes, one can find subdivisions that accomplish some added features and extend their usefulness. There are others that are not listed, but may be of use at some time and can be found in the technical literature. Casting of plastics is practiced to some degree in conjunction with some materials and products. Once the principles of shaping, which are presented in this chapter, are understood, it will not be difficult for the trained person to visualize to what degree tolerances will be a problem, the relative economics of production, tooling expenses, and other elements of manufacture that control the properties of a product.

10. Cost Estimating of Plastic Parts For Product Designers

When designing a plastic molded component, it is frequently desirable to compute an approximate cost in order to determine whether the considered design is within desired economical limits. The usual practice is to contact a molder, obtain a rough estimate, and proceed on the basis of this information to either redesign or finish the design on hand. This is time consuming, and on many occasions even this step is neglected because of the loss of time involved.

It is the intention of this chapter to provide the designer with the necessary guidelines that will enable him or her to approximate the cost factors to a degree that will point in the proper direction. The following is a simplified version of costing plastics. It covers the main and significant factors that include 85–90% of product value and should suffice for the requirements of a product designer when various designs are comparatively evaluated.

There are two basic ingredients that determine the value of a plastic part:

1. *The cost of the plastic resin (material)*

2. *The cost of molding*

The cost of the plastic material content is derived by calculating the cubic inches of the part and multiplying it by cost per cubic inch as shown in Fig. 5–1.

Once the cubic inches of the part are established, we will utilize these data as the basis for obtaining the remainder of the information, i.e., the cost of molding.

The following steps will lead to the molding cost:

(a) *The thickest portion* of part will determine the molding time required. Figure 10–1 shows how many pieces per hour can be molded of a part of a certain thickness in a single cavity. With a contemplated *yearly activity* and an ordering frequency of say *every 45 days*, we can tentatively decide on the *number of cavities* needed. The *number of cavities* should be such that it would take the 45-day requirement to produce in two hundred hours, or

$$\text{Number of cavities} = \frac{\text{45-day requirement}}{\text{200 hr} \times \text{pieces/hr}}$$

(b) *The number of cavities times the cubic inches per cavity* (or piece) leads us to the *machine size* by referring to Table 10–1.

There are a great number of cases in which the volume of material for all the cavities is small and the *projected areas* of the cavities are large; in that event, the choice of machine size is governed by the tonnage of the clamp required to keep the mold from opening during the injection of the material. The required tonnage is obtained by multiplying the *projected areas* of the cavities (parts) by

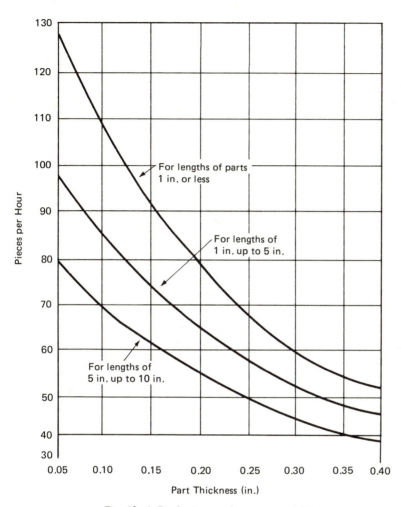

Fig. 10−1. Production per hour vs part thickness.

				TABLE 10−1					
Tons	50	75	100	125	150	200	250	300	350
Cubic inches	4.95	9.9	13.0	16.3	19.8	22.8	32.5	39.0	52.0
Cost per hour ($)	18	23	25	28	30	32	34	37	40
Tons	400	450	500	600	700	800	900	1000
Cubic inches	58.5	65.0	97.5	113.8	156.0	178.0	195.0	268
Cost per hour ($)	43	46	49	54	58	65	72	80

2−5 tons per square inch. The low value is for such materials as ABS, polystyrene, polyethylene, polypropylene, etc. The high value is for materials that are highly fluid when ready for injection, such as nylon and polycarbonate. Both the material volume and tonnage have to be satisfied while deciding on a machine size and, therefore, machine cost per hour. The larger press of the two requirements should prevail.

When machine size in terms of cubic inches and tonnage is decided upon, the *molding cost* can now be established by looking at Table 10−1. Molding cost per hour divided by pieces per hour gives molding cost per piece.

During the checking of cubic inches in the table against the press tonnage, only 70% of the machine volumetric capacity should be considered as the useful volume. The reason for this downgrading is that the heating capacity is based on polystyrene and most other materials require more energy for plasticating than does polystyrene. If the calculated volume for a part is more than 70% of rating, the selection calls for a higher tonnage press.

The costing rate per hour is based on conditions prevailing in 1981. Future corrections can be made about every three years by taking the average inflation rate during that period and increasing the table by the amount of such an average.

We now have the means of obtaining the *material cost* as well as the manufacturing cost per piece. The sum of these two multiplied by 1.2 gives the complete cost. The 20% addition is to cover other elements of expense. The description of the method of cost computing may give the impression of a lengthy procedure; however, the actual performance of the estimating will prove brief and will be worthwhile in doing.

A few examples will point to the practical interpretation of the above-outlined information.

Example No. 1: Polycarbonate Cover

Step A. The calculated cubic inch per piece is 1.6 in.^{3}. The material cost from Fig. 5−1 is 6.6 × 1.6 in.^{3} = 10.56¢ each.

Step B. The thickest portion is 0.110″ and length of part is $3^{13}/_{16}$″. From Fig. 10−1, the middle curve, we get 80 pieces/hr. The 45-day requirement is 60,000 pieces to be molded in 200 hr or

$$\frac{60,000}{200 \times 80 \text{ pieces/hr}} = 3.75 \text{ or } 4 \text{ } cavities$$

Step C. The number of cavities × in.^{3}/piece = 4 × 1.6 = 6.4 in.^{3} per operation.

Clamp Tonnage = $3^{13}/_{16}$ × $2^{3}/_{4}$ × 4 cavities × 5 tons/in.^{2}
                    = 210 tons

The 250-ton clamp press satisfies tonnage and cubic inch requirements. It's costing rate is $34 or

$$\frac{3400\cancel{c}}{4 \text{ cavities} \times 80 \text{ pieces/hr}} = \frac{3400}{320} = 10.625 \text{ or a } 10.625\cancel{c} \text{ manufacturing cost.}$$

Step D. Material cost of 10.56 + manufacturing cost 10.625 = 21.185 times 1.2 to cover other costs = 25.42¢ Total Cost.

Example No. 2: Nylon Switch Actuator

Step A. From Fig. 5–1, ¢/in.^{3} = 6.7, 0.475 in.^{3}/piece × 6.7 = 3.18¢/ piece.

Step B. Thickest portion is 0.165 × length of piece less than 1″, so upper curve gives 90 pieces/hr. The 45-day requirement is 30,000 pieces to be molded in 200 hr or

$$\frac{30,000}{200 \times 90 \text{ pieces/hr}} = 1.67 \text{ or } 2 \text{ cavities}$$

Step C. 2 × 0.475 in.^{3}/piece = 0.950 in.^{3} per operation. Tonnage = 2¾ × ½ × 2 (cavities) × 5 tons = 13.75 tons. A 50-ton press meets both requirements. Its costing rate is

$$\$18/\text{hr or} \frac{1800\cancel{c}}{2 \text{ cavities} \times 90 \text{ pieces/hr}} = \frac{1800}{180} \text{ or } 10\cancel{c} \text{ manufacturing cost.}$$

Step D. Material cost plus manufacturing cost = 3.18 + 10 = 13.18 each, times 1.2 = 15.18¢ each Total Cost.

Example No. 3: ABS Lid

Step A. 2.14 in.^{3}/piece × 3.08¢/in.^{3} = 6.59¢/pc.

Step B. Thickest portion 0.155 × 3½ length of piece, so middle curve will give 70 pieces/hr. The 45-day requirement is 20,000 pieces to be molded in 200 hr, or

$$\frac{20,000}{200 \times 70 \text{ pieces/hr}} = 1.43 \text{ or } 2 \text{ cavities}$$

Step C. 2 × 2.14 = 4.28 in.^{3} per operation or 75-ton press, since 4.28 in.^{3} is more than 70% of 4.95.

Tonnage 3½ × 3½ × 2 cavities × 3 tons = 73.5 ton

A 75-ton press will meet the cubic inch and tonnage requirements. Its costing rate is $23/hr or

$$\frac{2300}{2 \text{ cavities} \times 70 \text{ pieces/hr}} = \frac{2300}{140} \text{ or } 16.43\cancel{c} \text{ manufacturing cost.}$$

Step D. Material + manufacturing costs = 6.59 + 16.43 = 23.02 each times 1.2 = 27.62¢ each Total Cost.

For purposes of the above discussion, the curing or solidification time for all materials was figured as the same, although in practice some variation exists. For an evaluation of this type, this kind of an assumption is satisfactory, and the results should provide accurate guidance for design purposes.

Caution: Quoted prices may differ from approximate cost estimates because quotations will include not only profit, but will take into account such features as close tolerances and special performance requirements. The described cost estimates are intended for *comparative design evaluation.*

Glossary and Index

"A"-"B"-"C" stages: Steps in the reaction of thermosetting resins. In "A" the material is still soluble in certain liquids and is fusible. In "B" it softens when heated and swells when in contact with certain liquids, but does not fully fuse or dissolve. Molding compounds are usually in this stage. In "C" the resin is fully cured, i.e., it is relatively insoluble and infusible.

ablative plastics: Material with low thermal conductivity that absorbs or dissipates heat by pyrolysis (pyrolysis is decomposition of organic substances by heat). Used on exterior surfaces of temperature-sensitive structures to isolate them from hyperthermal effects of the environment.

abrasions and mar resistance: 42, 43.

ABS: 63–67

Absafil: Reinforced ABS, Fiberfil Division/ Dart Industries.

Abson: ABS, Abtec Chemical Co., ABS/PVC alloy.

absorption: (1) The penetration of the mass of one substance into another. (2) The process whereby energy is dissipated within a specimen placed in a field of radiation energy.

ABS-PC alloy: ABS–polycarbonate alloy: 64, 67

ABS-PVC alloy: ABS–Polyvinyl chloride: alloy 64, 67

accelerator: A substance that hastens a reaction, particularly one which speeds up the vulcanization of rubber. Also known as a *promoter*.

accumulator: (1) An auxiliary ram cylinder for fast delivery of plasticized melt. It is filled from the main barrel. Used on injection and blow-molding machines. (2) A container for storing hydraulic fluids under pressure and used for increased speed of hydraulic actuators.

acetal copolymer: 68, 69, 70, 71, 72

acetal homopolymer: 72, 73, 74, 75, 76, 77

acids: These substances are electrolytes. Their common characteristic is the formation of hydrogen ion in solution. An acid is also defined as a proton donor. Organic acids are characterized by the general formula for acid of the paraffin formula:

$$R-\overset{O}{\underset{C-O-H}{\|}}$$

R is an alkyl radical. The higher members of the group containing $12-18$ carbon atoms are called fatty acids. A carbon compound containing carboxyl groups, —COOH, indicated by the prefix carboxy, or the suffixes carboxylic, carbonyl, or simply -oic or -oyl.

Acrylafil: Reinforced SAN copolymer, Fiberfil Division, Dart Industries

acrylic: 78, 79, 80, 81

acrylic ester: An ester of acrylic acid or of a structural derivative of acrylic acid, e.g., methyl methacrylate.

Acrylite: Acrylic, CY/RO Ind.

acrylonitrile-butadiene-styrene (abbreviated ABS): 63

addition polymerization: 52

additives to polymers: 58
   flame retardance: 58
   UV light stability: 58
   heat stability: 58
   antioxidants: 58
   fillers and reinforcements: 58
   miscellaneous additives: 58

adiabatic: A process condition in which there is no gain or loss of heat from the environment.

aging: The chemical and/or physical changes that occur in a material after exposure to environmental conditions over a period of time. Can be natural or artificial (for testing) with specified conditions and predetermined time periods.

air-assist forming: A method of thermoforming in which air flow or air pressure is employed to preform partially the sheet immediately prior to the final pull-down onto the mold using vacuum.

air gap: In extrusion coating, the distance from the die opening to the nip formed by the pressure roll and the chill roll.

air ring: A circular manifold used to distribute an even flow of the cooling medium—air—onto a hollow tubular form passing through the center of the ring. In blown tubing, the air cools the tubing uniformly to provide uniform film thickness.

Alathon: Polyethelene, Du Pont Co.: 159

aldehydes: These substances are characterized by the presence of CHO radical. The simplest aldehyde is formaldehyde, HCHO.

aliphatic: These are organic compounds of the straight or branched chain arrangement of carbon atoms. The subgroups are (1) paraffins which are saturated and unreactive, (2) olefins which are unsaturated and quite reactive, and (3) acetylenes which contain a triple bond and are highly reactive.

alkalies: They are characterized by the negative radical, OH—known as the hydroxyl ion. Strong alkalies are often called caustic.

alkyd: 85–88

alkyl alcohols: Involves alhohols, *see* alkyl compounds.

alkyl compounds: Are derivatives in which one hydrogen atom has been replaced, such as, methyl, ethyl, propyl, butyl, amyl. The general formula is: C_nH_{2n+1}, where n is variable and the valence of the group is one.

alkyl halides: Involves chlorides or bromides.

allotropy: The existence of an element in more than one form in the same state (multiple existence), such as carbon in diamonds, in carbon black, and in graphite.

alloy: Composite material made up by blending polymers or copolymers with other polymers or elastomers under selected conditions, e.g., styrene–acrylonitrile copolymer resins blended with butadiene–acrylonitrile rubbers.

allyl–diallyl phthalate: 82

allyl diglycol carbonate (ADC): A peroxide (free radical) cured crystal-clear thermoset with exceptional scratch resistance. Used for optical purposes: 82

alpha cellulose (*see* cellulose)

alternating copolymer: 53

amides: Are derived from acids by a substitution of hydroxyl group OH with the NH_2 group.

amino: Indicates the (presence of an —NH_2 group.

aminos: Are nitrogen compounds of organic chemistry and are directly related to ammonia, NH_3, where one or more hydrogen atoms are replaced by organic groups. In amino compounds, the —NH_2 group replaces a hydrogen atom.

Amoco polyethylene: 159

Amoco polypropylene: 172

Amoco polystyrene: 180

amorphous: Means noncrystalline. Most plastics are amorphous at processing temperatures. Many of them retain this state even under solid conditions.

amorphous polymers: 56

anhydride: 66

anhydrous: Compounds from which water was removed; without water.

aniline ($C_6H_5NH_2$): An important organic base made by reacting chlorobenzene with aqueous ammonia in the presence of a catalyst. It is used in the production of aniline formaldehyde resins and in the manufacture of certain rubber accelerators and antioxidants.

aniline formaldehyde resins: Members of the aminoplastics family made by the condensation of formaldehyde and aniline in an acid solution.

The resins are thermoplastic and are used to a limited extent in the production of molded and laminated insulating materials. Products made from these resins have high dielectric strength and good chemical resistance.

anions: 47

annealing: A process of holding a material at a temperature near, but below, its melting point, the objective being to permit stress relaxation without distortion of shape. It is often used on molded articles to relieve stresses set up by flow into the mold.

anode: 46

antioxidant: Substance that prevents or slows down oxidation of material exposed to air.

antistatic agents: Methods of minimizing static electricity in plastics materials. Such agents are of two basic types: (1) metallic devices that come into contact with the plastics and conduct the static to earth. Such devices give complete neutralization at the time, but because they do not modify the surface of the material, it can become prone to further static during subsequent handling; (2) chemical additives that, mixed with the compound during processing, give a reasonable degree of protection to the finished products.

application of ABS: 63, 64, 65, 66

application, plastics, general considerations: 1, 2

aqueous solution: Solution in water.

arc resistance: time required for a given electrical current to render the surface of a material conductive because of carbonization by the arc flame.

Ardel: Polyarylate, Union Carbide Corp.: 153

aromatic compounds: 50

aromatic hydrocarbons: A compound of carbon and hydrogen with a closed saturated ring of carbon C atoms, e.g., benzene.

Arylon T: 138, 139

assembly, plastic parts: 246–253

ASTM designations: 61

ASTM specifications
D149: 29
D150: 30
D257: 31
D495: 32
D526: 17, 18
D570: 7
D621: 22
D635: 36
D638: 9–14
D648: 22, 23
D671: 20
D696: 24
D746: 24, 25

atactic: A chain of molecules in which the position of the methyl groups is more or less random.
atoms: 46
atomic number: 46
autoclave: (1) Closed strong vessel for conducting chemical reactions under high pressure; (2) in low-pressure laminating, a round or cylindrical container in which heat and gas pressure can be applied to resin-impregnated paper or fabric positioned in layers over a mold.
average molecular weight (viscosity method): The molecular weight of polymeric materials determined by the viscosity of the polymer in solution at a specific temperature. This gives an average molecular weight of the molecular chains in the polymer independent of specific chain length. Falls between weight average and number average molecular weight.
Azdel: Reinforced polypropylene sheet, PPG Industries.

back pressure: The viscosity resistance of a material to continued flow when a mold is closing. In extrusion, the resistance to the forward flow of molten material.
backlash: Working clearance of gears: 240, 241, 242
baffle: A device used to restrict or divert the passage of fluid through a pipeline or channel. In hydraulic systems the device, which often consists of a disc with a small central perforation, restricts the flow of hydraulic fluid in a high-pressure line. A common location for the disc is in a joint in the line. When applied to molds, the term is indicative of a plug or similar device located in a stream or water channel in the mold and designed to divert and restrict the flow to a desired path.
bag molding: A method of applying pressure during bonding or molding, in which a flexible cover, usually in connection with a rigid die or mold, exerts pressure on the material being molded, through the application of air pressure or drawing of a vacuum.
Bakelite: Polyethlene, Union Carbide.
  ethylene copolymers
  ethylene vinyl acetate copolymers

balanced equation: The chemical equation $Zn + O_2 \rightarrow ZnO$, if it is to be balanced according to the principle of matter conservation, should have exactly the same amount of material after the reaction as before. Therefore, since there are two atoms of oxygen on the left of equation, there must be two on the right, or two molecules of zinc oxide: $Zn + O_2 \rightarrow 2ZnO$. This now shows a shortage of one atom of Zn on the left; thus, the equation is $2Zn + O_2 = 2ZnO$.
Banbury: An apparatus for compounding materials composed of a pair of contrarotating rotors which masticate the materials to form a homogeneous blend. This is an internal-type mixer, which produces excellent mixing.
Barex: Acrylonitrile copolymer, Vistron Corp.
base: Another name for alkali or caustic.
batch number of material: 62
bearings, thermoplastic design: 246, 247, 248
Beetle: Urea, American Cyanamid Co.: 89, 90, 91, 92
beta gage (or beta-ray gage): A gage consisting of two facing elements: a beta-ray-emitting source and a beta-ray detector. When a sheet material is passed between the elements, some of the beta-rays are absorbed, the percentage absorbed being a measure of the area density or the thickness of the sheet.
bisphenol: Twice phenol.
blanking: The cutting of flat sheet stock to shape by striking it sharply with a punch while it is supported on a mating die. Punch presses are used. Also called *die cutting*.
bleed: To give up color when in contact with water or a solvent; undesired movement of certain materials in a plastic (e.g., plasticizers in vinyl) to the surface of the finished article or into adjacent material. Also called *migration*.
blister: A raised area on the surface of a molding caused by the pressure of the gases inside it on its incompletely hardened surface.
block copolymer: 53
blocking: An undesired adhesion between touching layers of a material, such as occurs under moderate pressure during storage or use.
blowing agents (*see* foaming agents)
blow molding: 258
blown tubing: A thermoplastic film which is produced by extruding a tube, applying a slight internal pressure to the tube to expand it while still molten, and subsequent cooling to set the tube. The tube is then flattened through guides and wound up flat on rolls. The size of blown tubing is determined by the flat width in inches as wound rather than by the diameter as in the case of rigid types of tubing.

blow pressure: The air pressure used to form a hollow part by blow molding.

blow rate: The speed at which the air enters the parison during the blow molding cycle.

blowup ratio: In blow molding, the ratio of the mold cavity diameter to the parison diameter. In blown tubing (film) the ratio of the final tube diameter (before gusseting, if any) to the original die diameter.

blueing: A mold blemish in the form of a blue oxide film which occurs on the polished surface of a mold as a result of the use of abnormally high mold temperatures.

booster or intensifier: Uses a large volume of low-pressure liquid to produce a low volume at high pressure.

boss: Protuberance on a plastic part designed to add strength, to facilitate alignment during assembly, to provide for fastenings, etc.

branched: In molecular structure of polymers (as opposed to *linear*), refers to side chains attached to the main chain. Side chains may be long or short.

breakdown voltage: The voltage required, under specific conditions, to cause the failure of an insulating material. (*See* dielectric strength.)

breaker plate: A perforated plate located at the rear end of an extruder head. It often supports the screens that prevent foreign particles from entering the die.

breathing: The opening and closing of a mold to allow gases to escape early in the molding cycle. Also called *degassing*. When referring to plastic sheeting, "breathing" indicates permeability to air.

brittleness temperature: 24

bromide: A binary salt containing negative monovalent bromine, e.g., sodium bromide.

"B" stage (*see* "A"-"B"-"C" stage)

bubbler mold cooling (injection molding): A method of cooling an injection mold in which a stream of cooling liquid flows continuously into a cooling cavity equipped with a coolant outlet normally positioned at the end opposite the inlet. Uniform cooling can be achieved in this manner.

bulk density: The mass per unit volume of a molding powder as determined in a reasonably large volume. The recommended test method is ASTM D1182-54.

bulk factor: Ratio of the volume of loose molding powder to the volume of the same weight of resin after molding.

bushing (extrusion): The outer ring of any type of a circular tubing or pipe die which forms the outer surface of the tube or pipe.

butene: A gas used in synthetic materials manufacture.

butyrate: 93

butt-fusion: A method of joining pipe, sheet, or other similar forms of a thermoplastic resin wherein the ends of the two pieces to be joined are heated to the molten state and then rapidly pressed together to form a homogeneous bond.

calender: To prepare sheets of material by pressure between two or more counterrotating rolls. The machine performing this operation.

calcium carbonate: A natural mineral used as a filler in plastics.

caprolactam: A cyclic amide-type compound, containing six carbon atoms. When the ring is opened, caprolactam is polymerizable into a nylon resin known as type-6 nylon or polycaprolactam.

Capron: Nylon, Allied Chemical Corp.
   type 6
   copolymer

carbonate: A compound resulting from the reaction (in plastics) of a carbonic acid with an organic compound; it forms an ester, e.g., diethyl carbonate, diphenyl carbonate.

carbon black: A black pigment produced by the incomplete burning of natural gas or oil. It is widely used as a filler, particularly in the rubber industry. Because it possesses useful ultraviolet protective properties, it is also much used in polyethylene compounds intended for such applications as cold water piping and black agricultural sheet.

carbonyl: The $C=O$ group itself. It is present in aldehydes, ketones, organic acids, sugars, and carboxyl groups.

carboxyl group: The chemical group characteristic of carboxylic acids, $COOH$ or CO_2H.

casein: A protein material precipitated from skimmed milk by the action of either rennet or dilute acid. Rennet casein finds its main application in the manufacture of plastics. Acid casein is a raw material used in a number of industries including the manufacture of adhesives.

cast: (1) To form a "plastic" object by pouring a fluid monomer–polymer solution into an open mold where it finishes polymerizing. (2) Forming plastic film and sheet by pouring the liquid resin onto a moving belt or by precipitation in a chemical bath.

catalyst: A substance which markedly speeds up the cure of a compound when added in small quantities as compared to the amounts of primary reactants. (*see also* hardener, inhibitor, promoter.)

cathode: 48

cations: 48
caustic: Strong alkalies are often called caustic. (*See* alkalies.)
cavity: Depression in a mold made by casting, machining, or hobbing or by a combination of these methods; depending on number of such depressions, molds are designated as *single cavity* or *multicavity*.
cavity filling with variation of wall: 216
Celanese nylon: 115
Celanex: Thermoplastic polyester, PBT, Celanese Plastics Materials Co.: 147
Celcon: Acetal, Celanese Plastics Materials Co.: 68–72
cellular plastics (*see* foamed plastics)
cellular striation: A layer of cells within a cellular plastic that differs greatly from the characteristic cell structure of the material.
celluloid: A thermoplastic material made by the intimate blending of cellulose nitrate with camphor. Alcohol is normally employed as a volatile solvent to assist plasticization and is subsequently removed.
cellulose: A natural high-polymeric carbohydrate found in most plants; the main constituent of dried woods, jute, flax, hemp, ramie, etc. Cotton is almost pure cellulose.
cellulose acetate: 93
cellulose acetate butyrate: 93
cellulose ester: A derivative of cellulose in which the free hydroxyl groups attached to the cellulose chain have been replaced wholly or in part by acidic groups, e.g., nitrate, acetate, or stearate groups. Esterification is effected by the use of a mixture of an acid with its anhydride in the presence of a catalyst, such as sulfuric acid. Mixed esters of cellulose, e.g., cellulose acetate butyrate, are prepared by the use of mixed acids and mixed anhydrides. Esters and mixed esters, a wide range of which are known, differ in their compatibility with plasticizers, in molding properties, and in physical characteristics. These esters and mixed esters are used in the manufacture of thermoplastic molding compositions.
cellulose nitrate (nitrocellulose): A nitric acid ester of cellulose manufactured by the action of a mixture of sulfuric acid and nitric acid on cellulose, such as purified cotton linters. The type of cellulose nitrate used for celluloid manufacture usually contains 10.8–11.1% nitrogen. The latter figure is the nitrogen content of the dinitrate.
cellulose plastics: 93–95
cellulose propionate: 93
cellulose triacetate: A cellulosic material made by reacting purified cellulose with acetic anhydride in the presence of a catalyst. It is used in

the form of film and fibers. Films and sheet are cast from clear solutions onto "drums" with highly polished surfaces. The film, which is of excellent clarity, has high tensile strength and good heat resistance and dimensional stability. Applications include book jackets, magnetic recording tapes, and various types of packaging. Cellulose triacetate sheet has somewhat similar properties to those of the film and is used to make such articles as safety goggles, map wallets, and transparent covers of many kinds.
center gated mold: An injection mold wherein the cavity is filled with resin through an orifice interconnecting the nozzle and the center of the cavity area. Normally, this orifice is located at the bottom of the cavity when forming items such as containers, tumblers, bowls, etc.
centipoise: A unit of viscosity, conveniently defined as the viscosity of water at room temperature. The following are approximate viscosities at room temperature in centipoise: water, 1; kerosene, 10; motor oil SAE 10, 100; caster oil, glycerin, 1000; corn syrup, 10,000; molasses, 100,000.
centrifugal casting: A method of forming thermoplastic resins in which the granular resin is placed in a rotatable container, heated to a molten condition by the transfer of heat through the walls of the container, and rotated so that the centrifugal force induced will force the molten resin to conform to the configuration of the interior surface of the container. Used to fabricate large diameter pipes and similar cylindrical items.
characteristics, plastics, general: 1, 2
charge: The measurement or weight of material used to load a mold at one time or during one cycle.
chase: An enclosure of any shape, used to (1) shrink-fit parts of a mold cavity in place; (2) prevent spreading or distortion in hobbing; (3) enclose an assembly of two or more parts of a split cavity block.
chemical compound: A substance containing two or more elements chemically united.
chemical equation: Shows what happens in a chemical reaction; $Zn + O_2 \rightarrow ZnO$—zinc plus oxygen gives zinc oxide. (*See* balanced equation.)
chemically foamed plastic: A cellular plastic whose structure is produced by gases generated from the chemical interaction of its constituents, e.g., polyurethane.
chemical resistance—plastics: 2, 61
chill roll: A cored roll, usually temperature controlled with circulating water, which cools the web before winding. For chill roll (cast) film, the surface of the roll is highly polished. In

extrusion coating, either a polished or a matte surface may be used depending on the surface desired on the finished coating.

chill roll extrusion (or cast fill extrusion): The extruded film is cooled while being drawn around two or more highly polished chill rolls cored for water cooling for exact temperature control.

chloride: A salt containing the Cl⁻ ion, usually a binary compound in which chlorine is the negative constituent.

chlorinated hydrocarbons: Hydrocarbon halogenated. A hydrocarbon in which one or more of the hydrogen atoms have been replaced by fluorine, chlorine, bromine, or iodine.

chromium plating: An electrolytic process that deposits a hard film of chromium metal onto working surfaces of other metals where resistance to corrosion, abrasion, and/or erosion is needed.

CIL (flow test): A method of determining the rheology or flow properties of thermoplastic resins; it was developed by Canadian Industries Limited. In this test, the amount of the molten resin that is forced through a specified size orifice per unit time when a specified, variable force is applied gives a relative indication of the flow properties of various resins.

clamping pressure: In injection molding and in transfer molding, the pressure that is applied to the mold to keep it closed, in opposition to the fluid pressure of the compressed molding material.

clay fillers: Used extensively in thermoset resins as a low cost extender to reduce resin cost in the molded product. The mineral kaolinite is the source for the clay filler called kaolin, a hydrous alumino silicate mineral. Kaolin is used as a viscosity-modifying filler in liquid polyesters for fiberglass molding. The increased viscosity tends to hold the glass fibers in the resin and eliminate show-through at the surface of the molded part. Very fine meshes of kaolin are used in thermoplastics.

clearance, working, backlash: 240, 241, 242

closed-cell foam: A cellular plastic in which there is a predominance of noninterconnecting cells.

coating weight: The weight of coating per unit area. In the United States usually "per ream," i.e., 500 sheets 24 in. × 36 in.(3000 ft²[61cm × 91cm(279 m²)]) but sometimes 1000 ft² (93 m²).

coefficient of expansion: 24

coefficient of friction, plastics: 2

cold flow: 33

cold molding: A procedure in which a composition is shaped at room temperature and cured by subsequent baking.

cold slug: The first material to enter an injection mold; so-called because in passing through a sprue orifice it is cooled below the effective molding temperature.

cold slug well: Space provided directly opposite the sprue opening in an injection mold to trap the cold slug.

cold stretch: Pulling operation, usually on extruded filaments, to improve tensile properties.

color, plastics, general considerations: 1

Co-Mer 7: Nylon type 6, Nypel, Inc.

compression mold: A mold that is open when the material is introduced and that shapes the material by heat and by the pressure of closing.

compression molding: 256

compression ratio: In an extruder screw, the ratio of volume available in the first flight at the hopper to the last flight at the end of the screw.

compressive strength: 16

concentrated solution: When a solution contains a relatively large amount of solute.

condensation polymerization: 53

conditioning procedure for ASTM: 44, 45

contact pressure resins: Liquid resins that thicken or resinify on heating and, when used for bonding laminates, require little or no pressure.

continuous phase: In a solution or mixture, the major component is called the continuous or external phase and the minor component is called the dispersed or internal phase. The latter may or may not be uniformly dispersed in the continuous phase.

cooling channels: Channels or passageways located within the body of a mold through which a cooling medium can be circulated to control temperature on the mold surface.

cooling fixture: Block of metal or wood holding the shape of a molded piece that is used to maintain the proper shape or dimensional accuracy of a molding after it is removed from the mold until it is cool enough to retain its shape without further appreciable distortion. Also known as *shrink fixture.*

copolymer: 53, 54

copolymerization methods
alternating copolymers: 53
block copolymers: 53
graft copolymers: 54
random copolymers: 53

core: (1) The central member of a sandwich construction (can be honeycomb material, foamed plastic, or solid sheet) to which the faces of the sandwich are attached; the central member of a plywood assembly. (2) A channel in a mold for circulation of heat-transfer media. (3) Part of a complex mold that molds undercut parts. Cores are usually withdrawn to one side before the main sections of the mold open. Also called *core pin.*

core drill: A device for making cooling channels in a mold.

corner
benefits of radiused: 207
sharp inside: 206, 207

corona resistance: A current passing through a conductor induces a surrounding electrostatic field. When voids exist in the insulation near the conductor, the high voltage electrostatic field may ionize and rapidly accelerate some of the air molecules in the void. These ions can then collide with the other molecules, ionizing them and thereby "eating" a hole in the insulation. Resistance to this process is called corona resistance.

corrosion resistance, plastics, general considerations: 1

cost, plastic material: 202–205

costing plastics for comparative designs: 260–264

covalence: 48

crazing: Fine cracks which may extend in a network on or under the surface or through a layer of a plastic material.

creep: 33–37, 62, 227, 228

creep data
background: 62, 227, 228
construction of curves: 62, 227, 228, 229
example: 228, 229
limitations for extrapolation: 227, 228, 229
predicting time of failure: 36, 227, 228

creep test, reading intervals: 36

criteria, plastic products, general: 1

crosshead (extrusion): A device generally employed in wire coating which is attached to the discharge end of the extruder cylinder. It is designed to facilitate extruding material at an angle. Normally, this is a 90° angle to the longitudinal axis of the screw.

cross-linking: Applied to polymer molecules, the setting-up of chemical links between the molecular chains. When extensive, as in most thermosetting resins, cross-linking makes one infusible supermolecule of all the chains.

cryogenics: (1) Study of the behavior of matter at temperatures below −328°F (−200°C). (2) The use of liquified gases.

crystalline: The self-arrangement of molecules in fixed patterns in solids. Crystallinity is a state of molecular structure in some resins that denotes uniformity and compactness of the molecular chains forming the polymer. Nor-

mally, it can be attributed to the formation of solid crystals having a definite geometric form.

crystalline polymers: 56

"C" stage (see "A"-"B"-"C" stage)

cull: Material remaining in a transfer chamber after mold has been filled. Unless there is a slight excess in the charge, the operator cannot be sure cavity is filled. Charge is generally regulated to control thickness of cull.

cure: To change the physical properties of a material by chemical reaction, which may be condensation, polymerization, or vulcanization; usually accomplished by the action of heat and catalysts, alone or in a combination, with or without pressure.

curing temperature: Temperature at which a cast, molded, or extruded product, a resin-impregnated reinforcing material, an adhesive, etc., is subjected to curing.

curing time (molding time): In the molding of thermosetting plastics, the interval of time between the instant of cessation of relative movement between the moving parts of a mold and the instant that pressure is released.

curling: A condition in which the parison curls upward and outward, sticking to the outer face of the die ring. Balancing the temperatures of the die and mandrel will normally relieve this problem.

curtain coating: A method of coating that may be employed with low viscosity resins or solutions, suspensions, or emulsions of resins in which the substrate to be coated is passed through and perpendicular to a freely falling liquid "curtain." The flow rate of the falling liquid and the linear speed of the substrate passing through the "curtain" are coordinated in accordance with the thickness of coating desired.

curvature: A condition in which the parison is not straight, but somewhat bending and shifting to one side, leading to a deviation from the vertical direction of extrusion. Centering of ring and mandrel can often relieve this defect.

cut-off: The line where the two halves of a compression mold come together; also called *flash groove* or *pinch-off.*

Cyanaprene: T P urethane, American Cyanamid Co.

cycle: The complete, repeating sequence of operations in a process or part of a process. In molding, the cycle time is the period, or elapsed time, between a certain point in one cycle and the same point in the next cycle.

cyclohexane dimethanol: A compound used to reduce reaction time in esterification.

Cycolac: ABS, Borg-Warner Chemicals: 66

Cycoloy: ABS/PC alloy, Borg-Warner Chemicals: 66
Cycovin: ABS/PVC alloy, Borg-Warner Chemicals: 66
Cymel: Melamine, American Cyanamid Co.: 91
Cyrolite: Acrylic copolymer, CY/RO Ind.
Cytor: Thermoplastic urethane, American Cyanamid Co.

Dacovin: PVC, rigid, Diamond Shamrock Corp.
dash-pot: A device used in hydraulic systems for damping down vibration. It consists of a piston attached to the part to be damped and fitted into a vessel containing fluid or air. It absorbs shocks by reducing the rate of change in the momentum of moving parts of machinery.
data sheet, blank: 61
daylight opening: Clearance between two platens of a press in the open position.
deckle rod: A small rod or similar device inserted at each end of the extrusion coating die that is used to adjust the length of the die opening.
decorative sheet: A laminated plastics sheet used for decorative purposes in which the color and/or surface pattern is an integral part of the sheet.
definite proportions law: The combination of elements in simple ratios—1:1, 2:1, 2:3, 3:4, etc.
deflashing: Covers the range of finishing techniques used to remove the flash (excess, unwanted material) on a plastic molding.
deflection temperature: 22, 23, 24
deformation under load: 22
degassing (*see* breathing)
degradation, properties: 206
degree of polymerization (DP): The number of structural units or mers in the "average" polymer molecule in a particular sample. In most plastics the DP must reach several thousand if worthwhile physical properties are to be obtained.
delamination: The separation of the layers in a laminate caused by the failure of the adhesive.
deliquescent: Capable of attracting moisture from the air.
Delrin: Acetal, Du Pont Co.: 72–77
denier: The weight (in grams) of 9000 m of synthetic fiber in the form of continuous filament.
density (*see also* specific gravity): Weight per unit volume of a substance, expressed in grams per cubic centimeter, pounds per cubic foot, etc.: 6

desiccant: Substance that can be used for drying purposes because of its affinity for water.
design
bearings, thermoplastic: 246, 247, 248
common features: 206–223
factors for bearings: 247,248
features—"good" and "poor": 210
plastic gears: 233–246
plastics approach: 224, 225
plastics, basic information: 225
plastics: 224–230
product features: 206
structural member: 230–233
destaticization: Treating plastics materials to minimize their accumulation of static electricity and, consequently, the amount of dust picked up by the plastics because of such charges.
detergents: Substances with a high surface activity. Similar, in general, to soaps; made synthetically to a large extent.
detraction, properties: 206
diacid: Two or double acid.
diamine: Polymerization catalyst.
diaphragm gate: Used in molding annular or tubular articles.
diatomic: Containing two atoms.
dibasic: Acids having two displaceable hydrogen atoms per molecule.
die adapter: The part of an extrusion die that holds the die block.
die blades: Deformable member(s) attached to a die body that determine the slot opening and that are adjusted to produce uniform thickness across the film or sheet produced.
die cutting: (1) Blanking. (2) Cutting shapes from sheet stock by striking it sharply with a shaped knife edge known as a "steel rule die." *Clicking* and *dinking* are other names for die cutting of this kind.
die gap: The distance between the metal faces forming the die opening.
dielectric: Insulating material. In radio-frequency preheating, dielectric may refer specifically to the material that is being heated.
dielectric constant: 30
dielectric heating (electronic heating): The plastic to be heated forms the dielectric of a condenser to which is applied a high-frequency (20–80 MHz) voltage. Dielectric loss in the material is the basis for this heating. Process used for sealing vinyl films and preheating thermoset molding compounds.
dielectric strength: 29
diffusion: Mixing or self-dispersion of gases without outside influence.
difunctional: Dual functions.
dilute solution: When a solution contains a relatively small amount of solute.

dimensional stability: Ability of a plastic part to retain the precise shape in which it was molded, fabricated, or cast.

dimer: A substance (comprising molecules) formed from two molecules of a monomer.

dip coating: Applying a plastic coating by dipping the article to be coated into a tank of melted resin or plastisol, then chilling the adhering melt.

dispersion: Finely divided particles of a material in suspension in another substance.

dissipation factor: 30

divalent and trivalent: Valence of 2 or 3.

doctor roll (doctor bar, doctor blade): A device for regulating the amount of liquid material on the rollers of a spreader.

double-shot molding: A means of turning out two-color parts in thermoplastic materials by successive molding operations.

dowel pin: Used for maintaining alignment between two or more parts of a plastic mold.

draft angle, types: 214

draft on protruding sections: 214, 215

drape forming: Method of forming thermoplastic sheet in which the sheet is clamped into a movable frame, heated, and draped over high points of a male mold. Vacuum is then pulled to complete the forming operation.

draw down ratio: The ratio of the thickness of the die opening to the final thickness of the product.

drawing: The process of stretching a therm plastic sheet or rod to reduce its cross-sectional area.

dry coloring: Method commonly used by fabricators for coloring plastics by tumble blending uncolored particles of the plastic material with selected dyes and pigments.

ductility: The extent to which a solid material can be drawn into a thinner cross section.

Dural: PVC, rigid, Alpha Chemical Corp.

Durez: 129

dwell: A pause in the application of pressure to a mold, made just before the mold is completely closed, to allow the escape of gas from the molding material.

dyes: Synthetic or natural organic chemicals that are soluble in most common solvents. Characterized by good transparency, high tinctorial strength, and low specific gravity.

Dylan: Ethylene vinyl acetate copolymer, ARCO/Polymers, polyethylene type I.

Dylark: Polystyrene copolymer, ARCO/Polymers.

Dyline: Polystyrene, GP, ARCO/Polymers, polystyrene, impact.

Dylite: Polystyrene, expandable bead, ARCO/Polymers.

Eccomold: Epoxy, Emerson and Cuming, Inc.

efflorescence: An occurrence when some salts lose their water of hydration upon exposure to the atmosphere.

ejector pin: A pin or thin plate that is driven into a mold cavity from the rear as the mold opens, forcing out the finished piece. Also called a *knockout pin.*

ejector return pins: Projections that push the ejector assembly back as the mold closes; also called *surface pins* and *return pins.*

Ekkcel: Aromatic copolyester, Dart Ind.

elastic limit: 13

elastic deformation: The part of the deformation of an object under load that is recoverable when the load is removed.

elastomers: 197, 198

electrode: 47

electroformed molds: A mold made by electroplating metal on the reverse pattern on the cavity. Molten steel may be then sprayed on the back of the mold to increase its strength.

electrolyte: A substance that will provide ionic conductivity when dissolved or in contact with water.

electron: 46

electronic treating: A method of oxidizing film of polyethylene to render it printable by passing the film between the electrodes and subjecting it to a high-voltage corona discharge.

electron valence: The combining power of atoms in polar compounds.

elongation: The fractional increase in length of a material stressed in tension: 227

elongation (*see* strain)

elongation under load with time: 227, 228

embossing: Techniques used to create depressions of a specific pattern in plastics film and sheeting.

emulsion: A suspension of fine droplets of one liquid in another.

emulsion polymerization: 52

encapsulating: Enclosing an article (usually an electronic component or the like) in a closed envelope of plastic, by immersing the object in a casting resin and allowing the resin to polymerize or, if hot, to cool.

endings

  -al: For names of aldehydes, example, chloral.

  -ene: Intended to show a type of compound or its derivatives in which there is a double bond linking two carbon atoms as one form of unsaturation.

  -ol: In a name indicates that the compound contains a hydroxy group.

  -one: Is used for ketones.

engineering plastics: Plastics that are most

frequently applied to engineering designs, e.g., gears, structural members, and technical components.

environmental stress cracking (ESC): The susceptibility of a thermoplastic article to crack or craze formation under the influence of certain chemicals and stress.

Epoxsilrub: Epoxy, Isochem Resins Co.

epoxy resins: 96−98

equilibrium with the solid phase: The solution will dissolve no more of the solid material.

Estane: Thermoplastic urethane, B. F. Goodrich.

ester: The reaction product of an alcohol and an acid.

ethers: Oxides with hydrocarbon radicals such as methyl ether. In those, two alkyl groups (alike or unlike) are connected through the oxygen atom: $R-O-R'$.

Ethofil: Reinforced polyethylene type III, Fiberfil Division, Dart Ind.

ethylene
 chlortrifluoroethylene: 106
 tetrafluoroethylene: 106

ethylene−vinyl acetate: Copolymers from these two monomers form a new class of plastic materials. They retain many of the properties of polyethylene, but have considerably increased flexibility and also have their density−elongation and impact resistance increased.

exotherm: (1) The temperature/time curve of a chemical reaction giving off heat, particularly the polymerization of casting resins. (2) The amount of heat given off.

expanded plastics (*see* foamed plastics)

expansion, thermal: 24

extender: A substance, generally having some adhesive action, added to a plastic composition to reduce the amount of the primary resin required per unit area.

extrudate: The product or material delivered by an extruder, such as film, pipe, wire coating, etc.

extrusion: 257, 258

extrusion coating: The resin is coated on a substrate by extruding a thin film of molten resin and pressing it onto or into the substrates or both, without the use of an adhesive.

fabricate: To work a material into a finished form by machining, forming, or other operation or to make flexible film or sheeting into end-products by sewing, cutting, sealing, or other operation.

fadeometer: An apparatus for determining the resistance of resins and other materials to fading. This apparatus accelerates the fading by subjecting the article to high-intensity ultraviolet rays of approximately the same wavelength as those found in sunlight.

family mold (injection): A multicavity mold wherein each of the cavities forms one of the component parts of the assembled finished object.

fatigue strength: 20

fatty acid: An organic acid obtained by the hydrolysis (saponification) of natural fats and oils, e.g., stearic and palmitic acids. These acids are monobasic, may or may not have some double bonds, and contain 16 or more carbon atoms.

features, product: 206

ferrite: A compound (a multiple oxide) of ferric oxide with another oxide such as sodium ferrite—$NaFeO_2$—more commonly a multiple oxide crystal.

fiber: This term usually refers to relatively short lengths of very small cross-sections of various materials. Fibers can be made by chopping filaments (converting). Staple fibers may be one-half to a few inches in length and usually 1 to 5 denier (0.5 to 1 mil in diameter in Marlex polyethylene).

Fibercore: Thermoset polyester bulk molding compound (BMC), Plumb Chemical Division.

Fiberite phenolic: 129, 130

filament: A variety of fiber characterized by extreme length, which permits its use in yarn with little or no twist and usually without the spinning operation required for fibers.

filament winding: Roving or single strands of glass, metal, or other reinforcements are wound in a predetermined pattern onto a suitable mandrel. The pattern is so designed as to give maximum strength in the directions required. The strands can either be run from a reel through a resin bath before winding or preimpregnated materials can be used. When the right number of layers have been applied, the wound mandrel is cured at room temperatures or in an oven.

fill-and-wipe: Parts are molded with depressed designs; after application of paint, the surplus is wiped off, leaving paint remaining only in depressed areas.

filler: A cheap, inert substance added to a plastic to make it less costly. Fillers may also improve physical properties, particularly hardness, stiffness, and impact strength. The particles are usually small, in contrast to those of reinforcements, but there is some overlap between the functions of the two.

fillet: A rounded filling of the internal angle between the two surfaces of a plastic molding.

film: An optional term for sheeting having a

nominal thickness not greater than 0.010 in. (0.025 mm).

fin: The web of material remaining in holes or openings in a molded part that must be removed in finishing.

fish eye: A fault in transparent or translucent plastics materials, such as film or sheet, appearing as a small globular mass and caused by incomplete blending of the mass with surrounding material.

flake: Used to denote the dry, unplasticized base of cellulosic plastics.

flame-retardant resin: A resin that is compounded with certain chemicals to reduce or eliminate its tendency to burn. For polyethylene and similar resins, chemicals such as antimony trioxide and chlorinated paraffins are useful.

flame spraying: Method of applying a plastic coating in which finely powdered fragments of the plastic, together with suitable fluxes, are projected through a cone of flame onto a surface.

flame treating: A method of rendering inert thermoplastic objects receptive to inks, lacquers, paints, adhesives, etc., in which the object is bathed in an open flame to promote oxidation of the surface of the article.

flammability
  ASTM: 26
  design requirements: 29
  UL 94: 26, 27, 28

flash: Extra plastic attached to a molding along the parting line; it must be removed before the part can be considered finished.

flash gate: A long, shallow rectangular gate.

flash mold: A mold designed to permit excess molding material to escape during closing.

flash point: The lowest temperature at which a combustible liquid will give off a flammable vapor that will burn momentarily.

flexible molds: Molds made of rubber or elastomeric plastics used for casting plastics. They can be stretched to remove cured pieces with undercuts.

flexural data, comparison: 16

flexural strength: 14
  relation to deflection: 15
  variable: 15

floating platen: A platen located between the main head and the press table in a multidaylight press and capable of being moved independently of them.

flock: Short fibers of cotton, etc., used as fillers for molding materials.

flow line (weld line): A mark on a molded piece made by the meeting of two flow fronts during molding.

flow marks: Wavy surface appearance of an object molded from thermoplastic resins caused by improper flow of the resin into the mold.

fluidized-bed coating: A method of applying a coating of a thermoplastic resin to an article in which the heated article is immersed in a dense-phase fluidized bed of powdered resin and thereafter heated in an oven to provide a smooth, pin-hole-free coating.

fluorescent pigments: By absorbing unwanted wavelengths of light and converting them into light of desired wavelengths, these colors seem to possess an actual glow of their own.

fluorinated ethylenepropylene (FEP): 107

fluorine: (1) A tricyclic hydrocarbon, C_6H_4-$CH_2C_6H_4$, used in resinous products. (2) A most reactive nonmetallic element and forms fluorides.

fluoroplastics (formerly called fluorocarbons): 99–106

foamed plastics: Resins in sponge form. The sponge may be flexible or rigid, the cells closed or interconnected, the density anything from that of the solid parent resin down to, in some cases, 2 lb/ft^3 (32 kg/m^3). Compressive strength of rigid foams is fair, making them useful as core materials for sandwich structures. Both the flexible and rigid types are good heat barriers.

foaming agents: Chemicals added to plastics and rubbers that generate inert gases on heating, causing the resin to assume a cellular structure.

foil decorating: Molding paper, textile, or plastic foils printed with compatible inks directly into a plastic part so that the foil is visible below the surface of the part as integral decoration.

force plug: The portion of a mold that enters the cavity block and exerts pressure on the molding compound, designated as *top force* or *bottom force* by position in the assembly; also called *plunger, piston, core.*

Formaldafil: Reinforced acetal, Fiberfil Division, Dart Industries.

formaldehyde (HCHO): A colorless gas (usually employed as a solution in water) that possesses a suffocating, pungent odor. It is derived from the oxidation of methanol or low-boiling petroleum gases such as methane, ethane, propane, or butane. It is widely used in the production of phenol formaldehyde (phenolic), urea formaldehyde (urea), and melamine formaldehyde (melamine) resins.

Fortiflex: Ethylene copolymer, Soltex Polymer Corp.
  polyethylene type II
  type III
  type IV

Fosta: Nylon type 6, American Hoechst Corp.: 115

Fostafoam: Polystyrene, expandable bead, American Hoechst Corp.
Fostalite: Polystyrene, GP, American Hoechst Corp.
Fosta polystyrene: 180
Fostarene: Polystyrene, GP, American Hoechst Corp.
Fosta Tuf-Flex: Polystyrene, impact, American Hoechst Corp.
friction calendering: A process whereby an elastomeric compound is forced into the interstices of woven or cord fabrics while passing through the rolls of calender.
friction welding: A method of welding thermoplastic materials whereby the heat necessary to soften the components is provided by friction.
functionality: The ability of a compound to form covalent bonds. Compounds may be monofunctional, difunctional, trifunctional, polyfunctional, i.e., one, two, three, many functional groups participating in a reaction, respectively.
furan resins: 107
furfural resin: A dark-colored synthetic resin of the thermosetting variety obtained by the condensation of furfural with phenol or its homologs. It is used in the manufacture of molding materials, adhesives, and impregnating varnishes. Properties include high resistance to acids and alkalies.
fuse: In plastisol molding, to heat the plastisol to the temperature at which it becomes a single homogeneous phase. In this sense, *cure* is the same as *fuse*.

Gafite: Thermoplastic polyester, PBT, GAF Corporation.
gate: In injection and transfer molding, the orifice through which the melt enters the cavity. Sometimes the gate has the same cross section as the runner leading to it; often, it is severely restricted.
gate, location and size: 215, 216
gears
  design with plastics: 233–246
  factors for precision in plastics: 237, 238, 239, 240
  involute 20° spur: 234
  life factors: 246
  quality and strength of plastics: 236, 237, 238, 239, 240
  strength: 242–246
  types of plastic: 233, 234
  velocity of plastic: 236

gelation: Formation of infinitely large polymer networks in the reaction mixture.
Genal: Phenolic, General Electric Company: 129
Geon: PVC, Rigid, B. F. Goodrich Co.: 184
glass-bonded mica: A moldable thermoplastic having a glass binder and mica filler (ceramoplastic).
glass transition: The change in an amorphous polymer or in amorphous regions of a partially crystalline polymer from a viscous or rubbery condition to a hard and relatively brittle one (or from hard to viscous condition). This transition generally occurs over a relatively narrow temperature region and is similar to a solidification of a liquid to a glassy state; it is not a phase transition. Not only do hardness and brittleness undergo rapid changes in this temperature region, but other properties such as thermal expansibility and specific heat also change rapidly. This occurrence has been called a second-order transition, rubber transition, and rubbery transition.
glass transition temperature (TG): The temperature region in which the glass transition occurs (*see* glass transition). The measured value of the glass transition temperature depends to some extent on the test method.
glucose: A carbohydrate having the molecular formula $C_6H_{12}O_6$. It is the main constituent of starch and cellulose.
glycol: It is a term for dihydric alcohols, which are similar to glycerol used in many plastic materials.
graft copolymer: 54
granular structure: Nonuniform appearance of finished plastic material owing to retention of, or incomplete fusion of, particles of composition, either within the mass or on the surface.
Grilamid: Nylon, type 12, Emser Industries, Inc.
Grilon: Nylon, type 6, Emser Industries, Inc.
grit blasted: A surface treatment of a mold in which steel grit or sand materials are blown to the walls of the cavity to produce a roughened surface. Air escape from the mold is improved and a special appearance of the molded article is often obtained by this method.
guide pins: Devices that maintain proper alignment of force plug and cavity as mold closes.
gum: An amorphous substance or mixture which, at ordinary temperatures, is either a very viscous liquid or a solid that softens gradually on heating and that either swells in water or is soluble in it. Natural gums, obtained from the cell walls of plants, are carbohydrates or carbohydrate derivatives of intermediate molecular weight.

gusset: A tuck placed in each side of a tube of blown tubing as produced to provide a convenient square or rectangular package, similar to that of the familiar brown paper bag or sack in subsequent packaging.

gutta percha: A rubberlike material obtained from the leaves and bark of certain tropical trees. Sometimes used for the insulation of electrical wiring and for transmission belting and various adhesives.

Halar: Fluoropolymer, ECTFE, Allied Chemical Corp.: 106

halide: A binary compound in the general type MX, MX_2, or MX_3 in which M is a metal and X is a halogen (F, Cl, Br, I). The halogens are the seventh group of the periodic table.

halocarbon plastics: Plastics based on resins made by the polymerization of monomers composed only of carbon and halogen.

halogens: These are a group of elements headed by fluorine in the vertical periodic table (chlorine, bromine, iodine). The word means "salt forming."

Halon: Fluoropolymer, TFE, Allied Chemical Corp.

hand mold: A mold taken out of the press after each shot for part removal.

hardener: A substance or mixture of substances added to plastic composition, or an adhesive to promote or control the curing reaction by taking part in it. The term is also used to designate a substance added to control the degree of hardness of the cured film. (*See also* catalyst.)

hardness, Rockwell: 21, 22

haze, transparent plastics: 8

head-forming for assembly: 249, 250

heat-distortion point: 22, 23, 24

heating chamber: In injection molding that part of the machine in which the cold feed is reduced to hot melt. Also called *heating cylinder.*

heat, permanent effect: 37

heat sealing: A method of joining plastic films by simultaneous application of heat and pressure to areas in contact. Heat may be supplied conductively or dielectrically.

heat treat: Term used to cover annealing, hardening, tempering, etc., of metals.

HF preheating (*see* dielectric heating)

high-load melt index (*see* melt index)

high polymer: A macromolecular substance that as indicated by the term "polymer" and by the name (e.g., polyvinyl chloride) and formula (e.g., CH_2CHCl) by which it is identified, consists of molecules which are (at least approximately) multiples of the low molecular unit.

high-pressure laminates: Laminates: molded and cured at pressures not lower than 1000 psi and more commonly in the range of 1200–2000 psi.

hinge, thermoplastic: 221

hob: A master model in hardened steel used to sink the shape of a mold into a soft steel block.

hobbing: Forming multiple mold cavities by forcing a hob into soft steel (or beryllium–copper) cavity blanks.

holes, molded blind: 219, 229

homologous: Closely related compounds.

homopolymer: 51, 52

hopper dryer: A combination feeding and drying device for extrusion and injection molding of thermoplastics. Hot air flows upward through the hopper containing the feed pellets.

hopper loader: A curved pipe through which molding powders are pneumatically conveyed from shipping drums to machine hoppers.

Hostalen: Polypropylene, American Hoechst Corp.

    polypropylene copolymers

    polyethylene type III

hot gas welding: A technique of joining thermoplastic materials (usually sheet) whereby the materials are softened by a jet of hot air from a welding torch and joined together at the softened points. Generally, a thin rod of the same material is used to fill and consolidate the gap.

hot-runner mold: A mold in which the runners are insulated from the chilled cavities and are kept hot. Parting line is at gate of cavity, and the runners are in separate plate(s), so they are not, as is the case usually, ejected with the piece.

hot stamping: Engraving operation for marking plastics in which roll leaf is stamped with heated metal dies onto the face of the plastics. Ink compounds can also be used. By means of felt rolls, ink is applied to type and by means of heat and pressure, type is impressed into the material, leaving the marking compound in the indentation.

hydrates: Compounds containing water that is combined in a definite proportion are called hydrates. The water present is water of hydration, also called *water of crystallization.*

hydrogen: The lightest element, with an atomic weight of about one.

hydrogenation: Chemical process whereby hydrogen is introduced into a compound. While oils do not readily mix with hydrogen, hydrogenation can be brought about using a catalyst, such as finely divided nickel or nickel alloy.

hydrolysis: Salts of strong acids and weak alkalies or of weak acids and strong alkalies, react with water itself. This is called hydrolysis; the salt is said to hydrolyze.

hydroxyl or hydroxy: It is a group of OH compounds (methyl alcohol, methanol, or carbinol). Alcohol is characterized by the general formula *ROH* where *R* represents an alkyl group.

hygroscopic: Tending to absorb moisture.

Hytrel: Thermoplastic elastomer, copolyester, Du Pont.

ide: Compounds ending with the suffix ide (sodium chloride, sodium sulfide, calcium chloride, aluminum oxide, etc.) contain only two elements.

imide: A nitrogen-containing acid having two double bonds.

immiscible: Descriptive of two or more fluids that are not mutually soluble.

impact resistance: 17

impact strength
application: 19, 20
Izod: 17
tensile: 18, 19
tests, evaluation: 19, 20

Implex: Acrylic, Rohm & Haas Co.

impregnation: The process of thoroughly soaking a material such as wood, paper, or fabric with synthetic resin so that the resin gets within the body of the material. The process is usually carried out in an impregnator.

impulse sealing: A heat-sealing technique in which a pulse of intense thermal energy is applied to the sealing area for a very short time, followed immediately by cooling. It is usually accomplished by using an RF-heated metal bar, which is cored for water cooling or is of such a mass that it will cool rapidly at ambient temperatures.

infrared: Part of the electromagnetic spectrum between the visible light range and the radar range. Radiant heat is in this range and infrared heaters are much used in sheet thermoforming.

inhibitor: A substance that slows down a chemical reaction. Inhibitors are sometimes used in certain types of monomers and resins to prolong storage life.

initiator: A substance that speeds up the polymerization of a monomer and becomes a component part of the chain.

injection blow molding: A blow molding process in which the parison to be blown is formed by injection molding.

injection molding: 254, 255, 256

injection molding cycle: The complete time cycle of operation utilized in injection molding of an object including injection, die close, and die open time.

injection pressure: The pressure on the face of the injection ram at which molding material is injected into a mold. It is usually expressed in psi.

injection ram: The ram that applies pressure to the plunger in the process of injection molding or transfer molding.

inorganic pigments: Natural or synthetic metallic oxides, sulfides, and other salts, calcined during processing at 1200–2100°F (650–1150°C). They are outstanding in heat and light stability, weather resistance, and migration resistance.

insert: An integral part of a plastics molding consisting of metal or other material that may be molded into position or may be pressed into the molding after the molding is completed.

inserts
advantageous fastening: 217
drawback to molding into part: 217
fastening: 216, 217, 218
molded in: 216, 217

inside corner interference: 206, 207

inside sharp corners: 206

insoluble: Substances that dissolve in water to only a very small extent.

insulation resistance: The electrical resistance of an insulating material to a direct voltage. It is determined by measuring the leakage of current that flows through the insulation.

insulation, thermal and electrical, general considerations: 1

interlock: A safety device designed to ensure that a piece of apparatus will not operate until certain precautions have been taken.

intermediate compounds: This means that the compound is in itself not useful or of little primary importance, but is used as a stepping stone for preparation of something else.

internal mixers: Mixing machines using the principle of cylindrical containers in which the materials are deformed by rotating blades or rotors. The containers and rotors are cored so that they can be heated or cooled to control the temperature of a batch. These mixers are extensively used in the compounding of plastics and rubber materials and have the inherent advantage of keeping dust and fume hazards to a minimum.

intrinsic viscosity: The intrinsic viscosity of a polymer is the limiting value at infinite dilution of the ratio of the specific viscosity of the polymer solution to its concentration in moles per liter:

$$n_1 = \lim_{C \to 0} \frac{n_{sp}}{C}$$

where n_1 = intrinsic viscosity, n_{sp} = specific viscosity, C = concentration in moles per liter. Intrinsic viscosity is usually estimated by determining the specific viscosity at several low concentrations and extrapolating the values of n_{sp}/C to $C = 0$. The concentration is expressed in terms of the repeating unit. In the case of polystyrene the repeating unit is $-CH_2-CH(C_6H_5)-$ and has a molecular weight of 104.

introfaction: The change in fluidity and wetting properties of an impregnating material, produced by the addition of an introfier.

introfier: A chemical that will convert a colloidal solution into a molecular one. (*See also* introfaction.)

ion exchange resins: Small granular or bead-like particles containing acidic or basic groups, which will trade ions with salts in solutions. Generally used for softening and purifying water.

ionic charge: 50

ionomer: 108–110

ionomer resins: 108–110

ions: 47

irradiation (atomic): As applied to plastics, refers to bombardment with a variety of subatomic particles, generally alpha, beta, or gamma rays. Atomic irradiation has been used to initiate polymerization and copolymerization of plastics and in some cases to bring about changes in the physical properties of a plastic material.

Isinglass: A white, tasteless gelatine derived from the bladder of fishes, usually the sturgeon. It is used as an adhesive and clarifying agent.

iso: A prefix meaning the same as in isomer (the same part), isotope (the same place). In plastics it means having a subordinate chain of one or more carbon atoms attached to a carbon of the straight chain.

Isochemrez: Epoxy, Isochem Resins Co.

isocyanate resins: Most applications for this resin are based on its combination with polyols (e.g., polyesters, polyethers, etc.). During this reaction, the reactants are joined through the formation of the urethane linkage, and, hence, this field of technology is generally known as urethane chemistry: 181, 182

isomers: These are different compounds with the same molecular formula. An example is given:

$$CH_3CH_2CH_2CH_2CH_3$$
n-pentane

$$CH_3$$
$$\diagdown$$
$$CHCH_2CH_3$$
$$\diagup$$
$$CH_3$$
dimethyl ethyl methane

$$CH_3$$
$$|$$
$$CH_3CCH_3$$
$$|$$
$$CH_3$$
tetramethyl methane

isotactic: A chain of unsymmetrical molecules combined head to tail with their methyl groups occupying the same relative position in space along the chain.

isotopes: Elements with the same atomic number are called isotopes.

ite and ate endings: Compounds with these endings contain oxygen in a negative radical. The ite represents the lower and the ate the higher proportion of oxygen (sodium nitrite, $NaNO_2$; sodium nitrate, $NaNO_3$).

Izod impact test: 17

jet molding: Processing technique characterized by the fact that most of the heat is applied as the material passes through a nozzle or jet, rather than in a heating cylinder as is done in conventional processes.

jet spinning: For most purposes similar to melt spinning. Hot gas jet spinning uses a directed blast or jet of hot gas to "pull" molten polymer from a die lip and extend it into fine fibers.

jetting: Turbulent flow of resin from an undersize gate or thin section into a thicker mold section, as opposed to laminar flow of material progressing radially from a gate to the extremities of the cavity.

jig: Tool for holding component parts of an assembly during the manufacturing process or for holding other tools. Also called a *fixture*.

joining plastics: 249, 253

joint, butt, edge, lap, scarf: Methods of joining material ends.

ketones: They resemble aldehydes to some extent, but also have properties of their own. The general formula:

Kel-F: Fluoropolymer, CTFE, 3M Co.: 105

Kinel: Polyimide, Rhone-Poulenc Polymers.

kirksite: An alloy of aluminum and zinc used for the construction of blow molds; it imparts a high degree of heat conductivity to the mold.

kiss-roll coating: This roll arrangement carries a metered film of coating to the web; at the line of web contact, it is split with part remaining on the roll, the remainder of the coating adhering to the web.

knife coating: A method of coating a substrate (usually paper or fabric) in which the substrate, in the form of a continuous moving web, is coated with a material whose thickness is controlled by an adjustable knife or bar set at a suitable angle to the substrate. In the plastics industry PVC formulations are widely used in this work, and curing is effected by passing the coated substrate into a special oven, usually heated by infrared lamps or convected air. There are a number of variations of this basic technique, and they vary according to the type of product required.

knit lines (*see* weld lines)

knock out pin (*see* ejector pin)

knot tenacity (knot strength): The tenacity in grams per denier of a yarn where an overhand knot is put into a filament or yarn being pulled to show the sensitivity to compressive or shearing forces.

Kodar PETG: 152

Kohinor: PVC, Rigid, Pantasote Co.

Kralastic: ABS, USS Chemicals.

Kraton: Thermoplastic elastomer, styrene block copolymr, Shell Chemical Co.

K-Resin: 180

Kynar: Fluoropolymer, PVDF, Pennwalt Corp.: 106

lacquer: Solution of natural or synthetic resins, etc., in readily evaporating solvents, which is used as a protective coating.

lactam: A cyclic amide produced from amino acids by the removal of one molecule of water.

laminar flow: Laminar flow of thermoplastic resins in a mold is accompanied by solidification of the layer in contact with the mold surface that acts as an insulating tube through which material flows to fill the remainder of the cavity. This type of flow is essential to duplication of the mold surface.

laminated plastics (synthetic resin-bonded laminate, laminate): 195, 196

laminated melamine cellulose: 196

laminated melamine glass fabric: 196

laminated phenolic cellulose paper: 196

laminated phenolic wood: 196

laminated thermosetting polyester glass matt: 196

land: (1) The horizontal bearing surface of a semipositive or flash mold by which excess material escapes. (*See* cut-off.) (2) The bearing surface along the top of the flights of a screw in a screw extruder. (3) The surface of an extrusion die parallel to the direction of melt flow.

latch: Device to hold together two members of a mold.

lay-up: (*n*) As used in reinforced plastics, the reinforcing material placed in position in the mold; also the resin-impregnated reinforcement. (*v*) The process of placing the reinforcing material in position in the mold.

***L/D* ratio:** A term used to define an extrusion screw, which denotes the ratio of the screw length to the screw diameter.

leach: To extract a soluble component from a mixture by the process of percolation.

leader pin (*see* guide pin)

leader pin bushing (*see* guide pin bushing)

Lexan: Polycarbonate, General Electric Co.: 142

light resistance: The ability of a plastics material to resist fading after exposure to sunlight or ultraviolet light. Nearly all plastics tend to darken under these conditions.

lignin plastics: Plastics based on resins made by the treatment of lignin with heat or by reaction with chemicals or with not more than an equal amount of synthetic resin. Used as binders or extenders.

linear molecule: A long-chain molecule as contrasted to one having many side chains or branches.

linters: Short fibers that adhere to the cotton seed after ginning. Used in rayon manufacture, as fillers in plastics, and as a base for manufacture of cellulosic plastics.

litharge: PbO. An oxide of lead used as an inorganic accelerator, as a vulcanizing agent for neoprene, and as an ingredient of paints.

LNP

filled nylons: 115

reinforced polyethylene: 159

reinforced polypropylene: 172

reinforced polystryrene: 180

reinforced PPS: 167

reinforced SAN: 188

reinforced sulfones: 191

load
  continuous: 225
  intermittent: 225
loading tray (charging tray): A device in the form of a specially designed tray that is used to load the charge simultaneously into each cavity of a multicavity mold by the withdrawal of a sliding bottom from the tray.
locating ring: A ring that serves to align the nozzle of an injection cylinder with the entrance of the sprue bushing and the mold to the platen of a molding machine.
loop tenacity (loop strength): The tenacity or strength value obtained by pulling two loops, as two links in a chain, against each other to demonstrate the susceptibility that a yarn, cord or rope has for cutting or crushing itself.
loss factor: The product of the power factor and the dielectric constant.
low-pressure laminates: In general, laminates molded and cured in the range of pressures from 400 psi down to and including pressure obtained by the mere contact of the plies.
Lucite: Acrylic, Du Pont Co.: 78–81
lug: An indentation or raised portion of the surface of a container, provided to control automatic (multicolor) decorating operations.
luminescent pigments: Special pigments available to produce striking effects in the dark. Basically there are two types: one is activated by ultraviolet radiation, producing very strong luminescence and, consequently, very eye-catching effects; the other type, known as phosphorescent pigment, does not require any separate source of radiation.
Lustran: ABS, Monsanto Co.
  SAN copolymer: 188
Lustrex: Polystyrene, GP, Monsanto Co.: 180
  polystyrene, impact

macerate: (*v*) To chop or shred fabric for use as a filler for a molding resin. (*n*) The molding compound obtained when so filled.
machine shot capacity: Refers to the maximum weight of thermoplastic resin that can be displaced or injected by the injection ram in a single stroke.
macromolecule: The large ("giant") molecules that make up the high polymer.
mandrel: (1) The core around which paper, fabric, or resin-impregnated fibrous glass is wound to form pipes or tubes. (2) In extrusion, the central finger of a pipe or tubing die.
manifold: A term used mainly with reference to blow molding and sometimes with injection

molding equipment. It refers to the distribution or piping system that takes the single channel flow output of the extruder or injection cylinder and divides it to feed several blow molding heads or injection nozzles.
mar and abrasion resistance: 42, 43
Marlex: Polyethylene, Phillips Chemical: 159
  type II
  type III
  type IV
Marlex, polypropylene: 173
masterbatch: A plastics compound that includes a high concentration of an additive or additives. Masterbatches are designed for use in appropriate quantities with the basic resin or mix so that the correct end concentration is achieved. For example, color masterbatches for a variety of plastics are extensively used since they provide a clean and convenient method of obtaining accurate color shades.
mat: A randomly distributed felt of glass fibers used in reinforced plastics lay-up molding.
matched metal molding: Method of molding reinforced plastics between two close-fitting metal molds mounted in a press.
material well: Space provided in a compression or transfer mold to care for bulk factor.
materials, plastic
  arrangement: 62
  choice of: 62
materials, list:
  ABS: 63–67
  acetal, copolymer: 68–72
  acetal, homopolymer: 72–77
  acrylic: 78–81
  alkyd (thermoset polyester): 85–88
  allyl-diallyl phthalate: 82–84
  amino (melamine and urea): 89–92
  cellulosic (acetate, butyrate, propionate): 93–95
  epoxy: 96–98
  fluoroplastics (Teflon, etc.): 99–106
  furan: 107
  industrial laminates: 195, 196
  ionomer: 108 to 110
  nitrile resin: 111
  nylon: 112–127
  phenolic: 128–131
  phenylene oxide (Noryl): 132–135
  poly(amide-imide) (Torlon): 136, 137
  polyaryl ether (Arylon T): 138, 139
  polybutylene: 140
  polycarbonate: 141–145
  polyester, aromatic: 153
  polyetherimide (Ultem): 154–157
  polyethylene: 158–160
  polyimide: 163–164
  polymethyl pentene: 165

melamine formaldehyde resin: 89
melt extractor: Usually refers to a type of injection machine torpedo, but could refer to any type of device that is placed in a plasticating system for the purpose of separating fully plasticated melt from partially molten pellets and material. It thus ensures a fully plasticated discharge of melt from the plasticating system.
melt fracture: An instability in the melt flow through a die starting at the entry to the die. It leads to surface irregularities on the finished article like a regular helix or irregularly spaced ripples.
melt index: The amount, in grams, of a thermoplastic resin that can be forced through a 0.0825-in.(2.1-mm) orifice when subjected to 2160 grams of force in 10 min at 190°C.
melt instability (*see* melt fracture)
melt strength: The strength of the plastic while in the molten state.
mer: 49
Merlon: Polycarbonate, Mobay Chemical Co.: 142
metallic pigments: A class of pigments consisting of thin opaque aluminum flakes or copper alloy flakes (known as bronze pigments). Incorporated into plastics, they produce unusual silvery and other metallike effects.
metallizing: Applying a thin coating of metal to a nonmetallic surface. May be done by chemical deposition or by exposing the surface to vaporized metal in a vacuum chamber or by metal powder spraying through a flame curtain.
metering screw: An extrusion screw that has a shallow constant depth and constant pitch section over, usually, the last three to four flights.
metathesis: When two compounds such as aluminum chloride and sodium hydroxide react by the simple interchange of radicals, forming aluminum hydroxide and sodium chloride, this is called double decomposition or metathesis.
methyl chloride (CH_3Cl): A compound used

in polymerization silicones and as a methylating agent in organic synthesis.
methyl methacrylate: 78, 111
methylpentene polymer: 165
MeV: Stands for million electron volts and is a unit of energy in motion of an electron.
Migralube nylon: 115, 118
migration of plasticizer: Loss of plasticizer from an elastomeric plastic compound with subsequent absorption by an adjacent medium of lower plasticizer concentration.
Minlon: Nylon, mineral-reinforced, Du Pont: 113
Mn (number-average molecular weight): The total weight of all molecules divided by the total number of molecules.
modified: Containing ingredients such as fillers, pigments, or other additives that help to vary the physical properties of a plastics material. An example is oil-modified resin.
modified polymers
  alloyed: 54
  polyblended: 54
modulus
  apparent: 35, 62
  creep: 35,62
  creep, apparent: 227
  of elasticity: 12, 13
  flexural: 15
  viscous: 35
moisture absorption, plastics: 7
moisture absorption, affect on product: 7
moisture vapor transmission: The rate at which water vapor permeates through a plastic film or wall at a specified temperature and relative humidity.
molar solution: A liquid containing a mole of substance per liter of solution (not per liter of water).
mold: (*v*) To shape plastic parts or finished articles by heat and pressure. (*n*) (1) The cavity or matrix into which the plastic composition is placed and from which it takes its form. (2) The assembly of all the parts that function collectively in the molding process.
mold release (*see* parting agent)
moldability of grades: 62
molding cycle: (1) The period of time occupied by the complete sequence of operations on a molding press requisite for the production of one set of moldings. (2) The operations necessary to produce a set of moldings without reference to the time taken.
molding powder: Plastic material in varying stages of granulation and comprising resin, filler, pigments, plasticizers, and other ingredients ready for use in the molding operation.
molding pressure: The pressure applied to

the ram of an injection machine or press to force the softened plastic completely to fill the mold cavities.

molding shrinkage (mold shrinkage, shrinkage, contraction) (*see* shrinkage)

mole: The molecular weight of a substance in grams.

molecular chain: 48, 49

molecular chain configurations: 55, 56

molecular weight distribution: The ratio of the weight-average molecular weight to the number-average molecular weight gives an indication of the distribution: 55

molecule: 47, 48, 227

molecules, method of presentation: 50

monoatomic: Containing single atoms.

monodispersity: Refers to a polymer system that is homogeneous in molecular weight, i. e., lacks molecular weight distribution.

monofilament (monofil): A single filament of indefinite length. Monofilaments are generally produced by extrusion. Their outstanding uses are in the fabrication of bristles, surgical sutures, fishing leaders, tennis-racquet strings, screen materials, ropes, and nets; the finer monofilaments are woven and knitted on textile machinery.

monomer: 49

Moplen: Polypropylene, Novamont Corp.

movable platen: The large back platen of an injection molding machine to which the back half of the mold is secured during operation. This platen is moved either by a hydraulic ram or a toggle mechanism.

multifilament yarn: The multifilament yarn is composed of a multitude of fine continuous filaments, often 5 to 100 individual filaments, usually with some twist in the yarn to facilitate handling. Multifilament yarn sizes are described in denier and range from 5–10 denier up to a few hundred denier. The large deniers, even in the thousands, are usually obtained by plying smaller yarns together. Individual filaments in a multifilament yarn are usually about 1–5 denier (which is about one-half mil to 1 mil diameter in Marlex polyethylene).

Mw (weight-average molecular weight): The sum of the total weights of molecules of each size multiplied by their respective weights divided by the total weight of all molecules.

Mylar: A film produced from the polyester of ethylene glycol and terephthalic acid. The fiber made by this method is called Dacron. Mylar and Dacron are trade names of Du Pont.

neck: In blow molding, the part of a container

where the shoulder cross-section area decreases to form the finish.

neck insert: Part of the mold assembly that forms the neck and finish. Sometimes called *neck ring.*

needle blow: A specific blow molding technique where the blowing air is injected into the hollow article through a sharpened hollow needle that pierces the parison.

neutralization: The reaction of acids and alkalies resulting in a neutral compound:

$$HCl + LiOH \rightarrow LiCl + H_2O$$

(hydrochloric acid plus lithium hydroxide gives lithium chloride plus water).

neutron: 46

Newtonian liquid: A liquid in which the rate of flow is directly proportional to the force applied. The viscosity is independent of the rate of shear; there is no yield value in Newtonian flow.

nip: The "V" formed where the pressure roll contacts the chill roll.

nitration: The substitution of the nitro group $-NO_2$ for hydrogen.

nitrile resin: 111

nitriles: Alkyl cyanides, and are thus called to be distinguished from inorganic cyanides.

Noan: Acrylic, modified, Richardson Co.

nondraft side: 214, 215

nonpolar: Having no concentrations of electrical charge on a molecular scale, thus, incapable of significant dielectric loss. Examples among resins are polystyrene and polyethylene: 48

nonrigid plastic: A nonrigid plastic is one which has a stiffness or apparent modulus of elasticity of not over 50,000 psi at 25°C (77°F) when determined according to ASTM test procedure D747-43T.

Noryl: Phenylene-oxide-based resin, General Electric Co.: 132, 133, 134, 135

notch sensitivity: 17, 112

notched impact strength: 206

Novimide: Epoxy, Isochem Resins Co.

Novolac: A phenolic-aldehyde resin which, unless one equivalent of methylene groups is added, remains permanently thermoplastic. (*See also* resinoid and thermoplastic.)

nozzle: The hollow cored metal nose screwed into the extrusion end of (a) the heating cylinder of an injection machine or (b) a transfer chamber where this is a separate structure. A nozzle is designed to form, under pressure, a seal between the heating cylinder or the transfer chamber and the mold. The front end of a nozzle may be either flat or spherical in shape.

nucleons: Protons and neutrons are called nucleons.

nucleus: 46

Ny-Kon: Reinforced nylon, LNP Corp.
P, type 6
R, type 66
Q, type 610
I, type 612
Nylafil: Reinforced nylon, Fiberfil Division, Dart Industries. Types 6, 66, 612.
Nylatron: Nylon, Polymer Corp. Type 66, 612.
Nylode: Reinforced nylon type 66, Fiberfil Division, Dart Industries.
nylon: 112–127

offset: A printing technique in which ink is transferred from a bath onto the raised surface of the printing plate by rollers. Subsequently, the printing plates transfer the ink to the object to be printed.
offset yield point: 12
oil-soluble resin: Resin at moderate temperatures will dissolve in, disperse in, or react with drying oils to give a homogeneous film of modified characteristics.
olefins: A group of unsaturated hydrocarbons of the general formula C_nH_{2n}, and named after the corresponding paraffins by the addition of "ene" or "ylene" to the stem. Examples are ethylene and propylene.
oleoresins: Semisolid mixtures of the resin and essential oil of the plant from which they exude and sometimes referred to as balsams. Oleoresinous materials also consist of products of drying oils and natural or synthetic resins.
omitted data in data sheets: 4
one shot molding: In the urethane foam field, indicates a system whereby the isocyanate, polyol, catalyst, and other additives are mixed together directly and a foam is produced immediately (as distinguished from *prepolymer*).
opaque: Descriptive of a material or substance that will not transmit light. Opposite of transparent. Materials that are neither opaque nor transparent are sometimes described as semiopaque, but are more properly classified as translucent.
open-cell foam: A cellular plastic in which there is a predominance of interconnected cells.
orange-peel: Said of injection moldings that have unintentionally rough surfaces.
organic: As used in connection with plastics is meant to cover compound of carbon and hydrogen—hydrocarbons and their derivatives.
organic pigments: Characterized by good brightness and brilliance. They are divided into insoluble organic toners and lake toners. The insoluble organic toners are usually free from salt-forming groups. Lake toners are practically pure, water-insoluble heavy metal salts of dyes without the fillers or substrates of ordinary lakes. Lakes, which are not as strong as lake toners, are water-insoluble heavy metal salts or other dye complexes precipitated upon or admixed with a base or filler.
organosol: A vinyl or nylon dispersion, the liquid phase of which contains one or more organic solvents. (*See also* plastisol.)
orientation: The alignment of the crystalline structure in polymeric materials so as to produce a highly uniform structure. Can be accomplished by cold drawing or stretching during fabrication.
orifice: The opening in the extruder die formed by the orifice bushing (ring) and mandrel.
orifice bushing: The outer part of the die in an extruder head.
ous and ic endings: "Ferrous" indicates a valence of two, whereas "ferric" indicates a valence of three.
overlay sheet (surfacing mat): A nonwoven fibrous mat (in glass, synthetic fiber, etc.) used as the top layer in a cloth or mat lay-up to provide a smoother finish or minimize the appearance of the fibrous pattern.
oxidation: The addition of oxygen to a compound or the reduction of hydrogen.
oxirane: A synonym for ethylene oxide. An oxirane group is one kind of epoxy group.
oxygen index: 26

paraffin: Also called alkane. A class of aliphatic hydrocarbons characterized by a straight or branched chain. It ranges in appearance from a gas (methane) to a waxy solid depending on molecular weight.
Paraplex: Thermoset polyester molding compound. Sheet molding compound (SMC). Premix.
parison: The hollow plastic tube from which a container, toy, etc., is blow molded.
parison swell: In blow molding, the ratio of the cross-sectional area of the parison to the cross-sectional area of the die opening.
part features in plastics, general considerations: 2
parting agent: A lubricant, often wax, used to coat a mold cavity to prevent the molded piece from sticking to it and thus to facilitate its removal from the mold. Also called *release agent*.
parting line: Mark on a molding or cast in where halves of the mold met in closing.
partitioned mold cooling: A large diameter

hole drilled into the mold (usually the core) and partitioned by a metal plate extending to near the bottom end of the channel. Water is introduced near the top of one side of the partition and removed on the other side. (*See* baffle.)

parylene: Poly-para-exylene, used in ultrathin films for capacitor dielectrics and as a pore-free coating. Films are formed by heating a monomer and condensing it on a cool surface.

Paxon: Polyethylene type III, Allied Chemical Corp.

PBT: 146–152

pearlescent pigments: A class of pigments consisting of particles that are essentially transparent crystals of a high refractive index. The optical effect is one of partial reflection from the two sides of each flake. When reflections from parallel plates reinforce each other, the result is a silvery luster. Effects possible range from brilliant highlighting to moderate enhancement of the normal surface gloss.

Pellethane: Thermoplastic urethane, Upjohn Co., CPR Division.

peptides: Low-molecular-weight polymers of amino acids. Arbitrarily designated as having molecular weights under 10,000. Higher-molecular-weight species are called polypeptides or proteins.

performance, product: 206

periodic table: 47

permanent set: The increase in length, expressed in a percentage of the original length, by which an elastic material fails to return to original length after being stressed for a standard period of time.

permeability, water vapor: 43, 44

Petrothene: Polyethylene, U.S. Industrial Chemicals Co.
　type I
　type II
　type III

pH: An expression of the degree of acidity or alkalinity of a substance. Neutrality is pH_7— acid solutions being under 7 and alkaline solutions over 7. pH meters are commercially available for accurate readings.

phenol: A class of aromatic organic compounds in which one or more hydroxy groups are attached directly to the benzene ring, i.e., phenol itself, the creosols, xylenols, resorcinols, naphthols.

phenolic resins: 128, 129, 130, 131

phenoxy resins: A high-molecular-weight thermoplastic polyester resin based on bisphenol-A and epichlorohydrin. Recently developed in the United States, the material is available in grades suitable for injection molding, extrusion, coatings, and adhesives.

phenylene oxide resin: 132, 133, 134, 135

phthalate esters: A main group of plasticizers produced by the direct action of alcohol on phthalic anhydride. The phthalates are the most widely used of all plasticizers and are generally characterized by moderate cost, good stability, and good all-round properties.

phthalocyanine pigments: Organic pigments of extremely stable chemical configuration resulting in very good fastness properties. These properties are enhanced by the formation of the copper complex, which is the phthalocyanine blue most used. The introduction of chlorine atoms into the molecule of blue gives the well-known phthalocyanine green, also usually in the form of copper complex.

pinch-off: A raised edge around the cavity in the mold that seals off the part and separates the excess material as the mold closes around the parison in the blow molding operation.

pinch-off land: The width of pinch-off blade that effects sealing of the parison.

pinch-off tail: The bottom of the parison that is pinched off when the mold closes.

pinhole: A very small hole in the extruded resin coating.

pinpoint gate: A restricted orifice of 0.030 in. (0.8 mm) or less in diameter through which molten resin flows into a mold cavity.

pitch: The distance from any point on the flight of a screw line to the corresponding point on an adjacent flight, measured parallel to the axis of the screw line or threading.

pitch, gear: 235, 236

Plaskon: Epoxy, Allied Chemical Corp.

Plaslode: Reinforced polyproplylene, Fiberfil Division, Dart Industries.

plastic: (*n*) One of many high-polymeric substances, including both natural and synthetic products, but excluding the rubbers. At some stage in its manufacture every plastic is capable of flowing, under heat and pressure, if necessary, into the desired final shape. (*adi*) Made of plastic; capable of flow under pressure or tensile stress.

plastic deformation: A change in dimensions of an object under load that is not recovered when the load is removed; opposed to elastic deformation.

plasticate: To soften by heating or kneading. Synonyms are: plastify, flux, and (imprecisely) plasticize.

plasticity: The quality of being able to be shaped by plastic flow.

plasticize: To soften a material and make it plastic or moldable, either by means of a plasticizer or the application of heat.

plasticizer: Chemical agent added to plastic

compositions to make them softer and more flexible.

plastics tooling: Tools, e.g., dies, jigs, fixtures, etc., for the metal-forming trades constructed of plastics, generally laminates or casting materials.

plastigel: A plastisol exhibiting gellike flow properties.

plastisols: Mixtures of resins and plasticizers that can be molded, cast, or converted to continuous films by the application of heat. If the mixtures contain volatile thinners also, they are known as *organosols.*

plastometer: An instrument for determining the flow properties of a thermoplastic resin by forcing the molten resin through a die or orifice of specific size at a specified temperature and pressure.

plate dispersion plug: Two perforated plates, held together with a connecting rod, that are placed in the nozzle of an injection molding machine to aid in dispersing a colorant in a resin as it flows through the orifices in the plates.

platens: The mounting plates of a press to which the entire mold assembly is bolted.

Plenco: Alkyd, Plastics Engineering Co.
  melamine
  melamine/phenolic
  phenolic

Plenco phenolics: 129

Plexiglas: Acrylic, Rohm & Haas Co.

plug forming: A thermoforming process in which a plug or male mold is used to partially preform the part before forming is completed using vacuum or pressure.

pneumatic: A system in which energy is transferred by compression, flow, and expansion of air.

pock marks: Irregular indentations on the surface of a blown container caused by insufficient contact of the blown parison with the mold surface. They are due to low blow pressure, air gas entrapment, or moisture condensation on mold surface.

poise: The unit of viscosity expressed as one dyne per second per square centimeter.

polar compound: 48

polishing roll(s): A roll or series of rolls, which have a highly polished chrome-plated surface, that are utilized to produce a smooth surface on sheet as it is extruded.

poly(amide-imide): 136, 137

polyacrylate: A thermoplastic resin made by the polymerization of an acrylic compound such as methyl methacrylate.

polyallomer: 174

polyamide: A polymer in which the structural units are linked by amide or thioamide groupings. Many polyamides are fiber-forming.

polyblends: 54

polybutylene: 140

poly/butylene terephthalate(PBT): 146–152

Polycarbafil: Reinforced polycarbonate, Fiberfil Division, Dart Industries.

Polycarbonate resins: 141, 142, 143, 144, 145

polyarylate: 153

polyaryl ether: 138, 139

polyester, aromatic: 153

polyester reinforced urethane: A poromeric material, which may have a urethane impregnation or a silicone coating for shoe uppers and industrial leathers. (Poromeric is microporosity.)

polyester, thermoplastic: 146–152

polyetherimide: 154–157

polyethersulfone: 193

polyethylene: 158

Poly-Eth: Polyethylene, Gulf Oil Chemicals
  polyethylene copolymers
  type I   type III
  type II   type IV

polyisobutylene: The polymerization product of isobutylene. It varies in consistency from a viscous liquid to a rubberlike solid with corresponding variation in molecular weight from 1000 to 400,000.

Polyflam: ABS, A. Schulman, Inc.
  polypropylene
  polypropylene copolymers
  polystyrene, impact

polyimide, thermoplastic: 163, 164

polyimide, thermoset: 163, 164

Polyman: ABS/PVC alloy, A. Schulman, Inc.
  ethylene copolymers and alloys

polymer
  formation: 46–54
  individual (*see* materials list)
  manufacturing tolerances: 62
  structure: 51, 52, 53

polymerization: 51

polymerization methods
  addition: 52
  bulk: 52
  condensation: 53
  emulsion: 52
  solution: 52
  suspension: 53

polymethyl methacrylate: A thermoplastic material composed of polymers of methyl methacrylate. It is a transparent solid with exceptional optical properties and good resistance to water. It is obtainable in the form of sheets, granules, solutions, and emulsions. It is extensively used for aircraft domes, lighting fixtures, decorative articles, etc.; it is also used in optical instruments and surgical appliances.

polymethyl pentene: 165

polyol: A polyhydric alcohol, i.e., one containing three or more hydroxyl groups. Those having three groups are glycerols; those with more than three are called sugar alcohols.

polyolefin (*see* olefin)

polyphenyl sulfone: 193

polyphenylene oxide: 132–138

polyphenylene sulfide: 166–170

polypropylene: 171–177

polysaccharides: Naturally occurring polymers that consist of simple sugars. Examples are starch and cellulose.

polyset: Epoxy, Morton Chemical Co.

polystyrene: 178–180

polysulfides: Polymers containing sulfur and carbon linkage, and example of which is thiokol rubber.

polysulfone: 193, 194

polyterpene resins: Thermoplastic resins obtained by the polymerization of turpentine in the presence of catalysts. These resins are used in the manufacture of adhesives, coatings, and varnishes and in food packaging. They are compatible with waxes, natural and synthetic rubbers, and polyethylene.

polytetrafluoroethylene (PTFE) resins: 105

polytetramethylene terephthalate: 146

polyurethane resins: 181, 182

polyvinyl acetal: A member of the family of vinyl plastics. Polyvinyl acetal is the general name for resins produced from a condensation of polyvinyl alcohol with an aldehyde. There are three main groups: polyvinyl acetal itself, polyvinyl butyral, and polyvinyl formal. Polyvinyl acetal resins are thermoplastics that can be processed by casting, extruding, molding, and coating, but their main uses are in adhesives, lacquers, coatings, and films.

polyvinyl acetate: A thermoplastic material composed of polymers of vinyl acetate in the form of a colorless solid. It is obtainable in the form of granules, solutions, latices, and pastes and is used extensively in adhesives for paper and fabric coatings and in bases for inks and lacquers.

polyvinyl alcohol: A thermoplastic material composed of polymers of the hypothetical vinyl alcohol. Usually a colorless solid, insoluble in most organic solvents and oils, but soluble in water when the content of hydroxy groups in the polymer is sufficiently high. The product is normally granular. It is obtained by the partial hydrolysis or by the complete hydrolysis of polyvinyl esters, usually by the complete hydrolysis of polyvinyl acetate. It is mainly used for adhesives and coatings.

polyvinyl butyral: A thermoplastic material derived from a polyvinyl ester in which some or all of the acid groups have been replaced by hydroxyl groups and some or all of these hydroxyl groups replaced by butyral groups by reaction with butyraldehyde. It is a colorless flexible tough solid. It is used primarily in interlayers for laminated safety glass.

polyvinyl carbazole: A thermoplastic resin, brown in color, obtained by reacting acetylene with carbazole. The resin has excellent electrical properties and good heat and chemical resistance. It is used as an impregnant for paper capacitors.

polyvinyl chloride acetate: A thermoplastic material composed of copolymers of vinyl chloride and vinyl acetate; a colorless solid with good resistance to water and concentrated acids and alkalies. It is obtainable in the form of granules, solutions, and emulsions. Compounded with plasticizers it yields a flexible material superior to rubber in aging properties. It is widely used for cable and wire coverings, in chemical plants, and in protective garments.

polyvinyl fluoride: 106

polyvinyl formal: One of the groups of polyvinyl acetal resins, made by the condensation of formaldehyde in the presence of polyvinyl alcohol. It is used mainly in combination with cresylic phenolics, for wire coatings, and for impregnations, but can also be molded, extruded, or cast. It is resistant to greases and oils.

polyvinyl chloride (PVC): 183, 184

polyvinylidene chloride: Saran of Dow Chemical. A thermoplastic material composed of polymers of vinylidene chloride (1,1-dichloroethylene). It is a white powder with softening temperature at 365–392°F (185–200°C). The material is also supplied as a copolymer with acrylonitrile or vinyl chloride, giving products that range from the soft flexible type to the rigid type.

polyvinylidene fluoride: 106

porous molds: Molds that are made up of bonded or fused aggregate (powdered metal, coarse pellets, etc.) in such a manner that resulting mass contains numerous open interstices of regular or irregular size through which either air or liquids may pass through the mass of the mold.

positive mold: A mold designed to trap all the molding material when it closes.

postforming: The forming, bending, or shaping of fully cured, C-stage thermoset laminates that have been heated to make them flexible. On cooling, the formed laminate retains the contours and shape of the mold over which it has been formed.

pot life (*see* working life)

potting: Similar to *encapsulating*, except that steps are taken to ensure complete penetration of all the voids in the object before the resin polymerizes.

power factor: 30

PPS: 166–170

precision factors for plastic gears: 237, 238

preform: (*n*) A compressed tablet or biscuit of plastic composition used for efficiency in handling and accuracy in weighing materials. (*v*) To make plastic molding powder into pellets or tablets.

preheat roll: In extrusion coating, a heated roll installed between the pressure roll and unwind roll whose purpose is to heat the substrate before it is coated.

preheating: The heating of a compound prior to molding or casting in order to facilitate the operation or to reduce the molding cycle.

premix: In reinforced plastic molding, the material made by "do-it-yourselfers," molders, or end-users who purchase polyester or phenolic resin, reinforcement, fillers, etc., separately and mix the reinforced molding compounds on their own premises.

preplastication: Technique of premelting injection molding powders in a separate chamber, then transferring the melt to the injection cylinder. Device used for preplastication is commonly known as a preplasticizer. (*See also* plasticate.)

preplasticizer (*see* preplastication)

prepolymer: A chemical intermediate whose molecular weight is between that of the monomer or monomers and the final polymer.

prepreg: A term generally used in reinforced plastics to mean the reinforcing material containing or combined with the full complement of resin before molding.

preprinting: In sheet thermoforming, the distorted printing of sheets before they are formed. During forming, the print assumes its proper proportions.

press polish: A finish for sheet stock produced by contact, under heat and pressure, with a very smooth metal that gives the plastic a high sheen.

pressure forming: A thermoforming process wherein pressure is used to push the sheet to be formed against the mold surface as opposed to using a vacuum to suck the sheet flat against the mold.

pressure imbalance in cavity: 216

pressure pads: Reinforcements of hardened steel distributed around the dead areas in the faces of a mold to help the land absorb the final pressure of closing without collapsing.

pressure-sensitive adhesive: An adhesive that develops maximum bonding power by applying only a light pressure.

primer: A coating applied to a surface prior to the application of a plastic coating, adhesive, lacquer, enamel, etc., to improve the performance of the bond.

price of plastic materials: 202–205

printed circuit: An electrical or electronic circuit produced mainly from copper-clad laminates.

processing consideration of filled and reinforced polymers: 60

Profax: Polypropylene, Hercules, Inc.: 173 polypropylene copolymers.

Profil: Reinforced polypropylene, Fiberfil Division, Dart Industries.

profile die: An extrusion die for the production of continuous shapes, excepting tubes and sheets.

printing of plastics: Methods of printing plastics materials, particularly thermoplastic film and sheet, have developed side by side with the growth of usage of the materials and are today an important part of finishing techniques. Basically, the printing processes used are the same as in other industries, but the adaptation of machinery and development of special inks have been a constant necessity, particularly as new plastics materials have arrived, each with its own problems of surface decoration. Among the printing processes commonly used are gravure, flexographic, inlay (or valley), and silk screen.

programming: The extrusion of a parison that differs in thickness in the length direction in order to equalize wall thickness of the blown container. It can be done with a pneumatic or hydraulic device that activates the mandrel shaft and adjusts the mandrel position during parison extrusion (parison programmer, controller, or variator). It can also be done by varying extrusion speed on accumulator-type blow molding machines.

projected area: The total projected area at the parting line of all moldings, runners, sprues, vents, or culls in a mold.

promoter: A chemical, itself a feeble catalyst, that greatly increases the activity of a given catalyst.

properties, plastics
    data sheet: 3, 4
    specific: 1, 2

properties, polymer, general: 56
property change by fillers and reinforcements: 59
property, degradation: 206–211
property detractors: 206–211
propionate: 93
proportional limit: 13
propyl compounds: Derivatives in which one hydrogen atom of propane has been replaced.
proton: 46
prototype mold: A simplified mold construction often made from a light metal casting alloy or from an epoxy resin in order to obtain information for the final mold and/or part design.
pulp: A form of cellulose obtained from wood or other vegetable matter by prolonged cooking with chemicals.
pulp molding: Process by which a resin-impregnated pulp material is preformed by application of a vacuum and, subsequently, oven cured or molded.
pultrusion: A process in which thermosetting resin and fiber reinforcements are combined and pulled through a die in contrast to being pushed or extruded as practiced in extrusion of thermoplastics.
purging: Cleaning one color or type of material from the cylinder of an injection molding machine or extruder by forcing it out with the new color or material to be used in subsequent production. Purging materials are also available.
PVC: 183, 184
pyrrones: Polyimidazo-pyrrolones synthesized from dianhydride and tetramines, soluble only in sulfuric acid, resistant to temperatures up to 1110°F (600°C).

quench (thermoplastics): A process of shock cooling thermoplastic materials from the molten state.
quench bath: The cooling medium used to quench molten thermoplastic materials to the solid state.
quench-tank extrusion: The extruded film is cooled in a quench-water bath.

Radel: Polyphenyl sulfone, Union Carbide Corp.: 190–192
radicals: When more than two atoms are present in a compound, some of them occasionally group themselves and behave as a unit in a chemical reaction. The SO_4 is called a sulfate radical; OH is a hydroxyl radical.

radio-frequency (rf) preheating: A method of preheating used for molding materials to facilitate the molding operation or reduce the molding cycle. The frequencies most commonly used are between 10 and 100 MHz.
radio-frequency welding: A method of welding thermoplastics using a radio-frequency field to apply the necessary heat. Also known as high-frequency welding.
ram (*see* force plug)
radius, benefits of inside corner: 206, 207
ram travel: The distance the injection ram moves in filling the mold, in either injection or transfer molding
random copolymer: 53
rates of speed, tensile strength: 11
rayon: The generic term for fibers, staple, and continuous filament yarns composed of regenerated cellulose, but also frequently used to describe fibers obtained from cellulose acetate or cellulose triacetate. Rayon fibers are similar in chemical structure to natural cellulose fibers (e.g., cotton) except that the synthetic fiber contains shorter polymer units. Most rayon is made by the viscose process.
ream: Usually 500 sheets, 24″ × 36″ of industrial paper. Sometimes expressed as 3000 ft².
reciprocating screw: An extruder system in which the screw when rotating is pushed backward by the molten polymer that collects in front of the screw. When sufficient material has been collected, the screw moves forward and forces the material through the head and die at a high speed.
reduction: Addition of electrons.
recycle: Ground material from flash and trimmings that after mixing with a certain amount of virgin material is fed back into the blow molding machine.
reflectance, luminous: 8
regenerated cellulose (cellophane): A transparent cellulose plastics material made by mixing cellulose xanthate with a dilute sodium hydroxide solution to form a viscose. Regeneration is carried out by extruding the viscose in sheet form into an acid bath to create regenerated cellulose. The material is very widely used as a packaging and overwrapping material of exceptional clarity. The film has also good electrical properties and is resistant to oils and greases. Included among recent applications is the use of the material as release agent in reinforced plastics moldings.
reinforced molding compound: Compound supplied by raw material producers in the form of ready-to-use materials, as distinguished from premix.
reinforcement: A strong inert material bound

into a plastic to improve its strength, stiffness, and impact resistance. Reinforcements are usually long fibers of glass, sisal, cotton, etc.— in woven or nonwoven form. To be effective, the reinforcing material must form a strong adhesive bond with the resin.

relative humidity: Ratio of the quantity of water vapor present in the air to the quantity that would saturate it at any given temperature.

relative viscosity: The relative viscosity of polymer in the solution is the ratio of the absolute viscosities of the solution (of stated concentration) and of the pure solvent at the same temperature.

$$n_r = n/n_0$$

where n_r = relative viscosity, n = absolute viscosity of polymer solution, n_0 = absolute viscosity of pure solvent.

release agent (*see* parting agent)

relief angle: The angle of the cutaway portion of the pinch-off blade measured from a line parallel to the pinch-off land.

resiliency: Ability to regain quickly an original shape after being strained or distorted.

resin: Any of a class of solid or semisolid organic products of natural or synthetic orgin, generally of high molecular weight with no definite melting point. Most resins are polymers.

resinoid: Any of the class of thermosetting synthetic resins, either in their initial temporarily fusible state or in their final infusible state.

resin pocket: An apparent accumulation of excess resin in a small, localized section visible on cut edges of molded surfaces. Most noticeable in filled resins. Also called resin separation.

resistance
  arc: 32
  electrical: 31

resistivity: The ability of material to resist passage of electrical current either through its bulk or on a surface. The unit of volume resistivity is the ohm-cm; of surface resistivity, the ohm.

restricted gate: A very small orifice between runner and cavity in an injection or transfer mold. When the piece is ejected, this gate breaks cleanly, simplifying separation of runner from piece.

restrictor ring: A ring-shaped part protruding from the torpedo surface that provides increase of pressure in the mold to improve, e.g., welding of two streams.

retainer plate: The plate on which demountable pieces, such as mold cavities, ejector pins, guide pins, and bushings are mounted during molding; usually drilled for steam or water.

return pins: Pins that return the ejector mechanism to molding position.

rheology: The study of flow of polymers.

ribs: 209–213

rigid PVC: Polyvinyl chloride or a polyvinyl chloride/acetate copolymer characterized by a relatively high degree of hardness; it may be formulated with or without a small percentage of plasticizer.

rigid resin: One having a modulus high enough to be of practical importance, e.g., 10,000 psi or greater in contrast to flexible resin.

RIM process: 255, 256

ring gate: Annular opening for entrance of material into cavity of injection or transfer molds. Used for symmetrical flow of material.

ring structure of compounds: 50

Rogers phenolic: 130

Rockwell hardness: 21, 22

roller coating: Used for applying paints to raised designs or letters.

roll mill: Two rolls placed in close relationship to one another used to admix a plastic material with other substances. The rolls turn at different speeds to produce a shearing action to the materials being compounded.

rosin: A resin obtained as a residue in the distillation of crude turpentine from the sap of the pine tree (gum rosin) or from an extract of the stumps and other parts of the tree (wood rosin).

rotational casting (or molding), rotomolding: 258, 259

Rotoflame: Polyethylene type II, Rototron Corp.

Rotothene: Polyethylene type I, Rototron Corp.
  type III

Rotothon: Polypropylene copolymers, Rototron Corp.

roving: A form of fibrous glass in which spun strands are woven to a tubular rope. The number of strands is variable, but 60 is usual. Chopped roving is commonly used in preforming.

Roylar: Thermoplastic urethane, Uniroyal, Inc.

RTV: Room-temperature vulcanized.

Rucoblend: PVC, rigid, Ruco Division, Hooker Chemicals and Plastics Corp.

Rumiten: Polyethylene, type I, Rumianca Chemical Corp.

runner: In an injection or transfer mold, the channel, usually circular, that connects the sprue with the gate to the cavity.

Rynite: Thermoplastic polyester, modified PET, Du Pont: 147

Ryton: Polyphenylene sulfide, Phillips Chemical Co.: 167

sag: The extension locally (often near the die

face) of the parison during extrusion by gravitational forces; this causes necking-down of the parison. Also refers to the flow of a molten sheet in a thermoforming operation.

salts: Result from reaction of acids and alkalies.

SAN: 187, 188, 189

sandwich constructions: Panels composed of a lightweight core material—honeycomb, foamed plastic, etc.—to which two relatively thin, dense, high-strength faces or skins are adhered.

sandwich heating: A method of heating a thermoplastic sheet prior to forming that consists of heating both sides of the sheet simultaneously.

saponification: A reaction of fats or glycerides with caustic soda and caustic potash.

saturated compounds: 50, 51

saturation: When the solution will dissolve no more of the solute and its excess remains solid.

scrap: Any product of a molding operation that is not part of the primary product. In compression molding, this includes flash, culls, and runners and is not reusable as a molding compound. Injection molding and extrusion scrap (runners, rejected parts, sprues, etc.) can usually be reground and remolded.

scratch resistance (*see also* abrasion and mar resistance): 42, 43

screw plasticating injection molding: A technique in which the plastic is converted from pellets to a viscous melt by means of an extruder screw that is an integral part of the molding machine. Machines are either single stage (in which plastication and injection are done by the same cylinder) or double stage (in which the material is plasticated in one cylinder and then fed to a second for injection into a mold).

screws for joining plastics: 249

sealing plane: The plane on the inside of a bottle cap along the sealing surface.

sealing surface: The surface of the finish of the container on which the closure forms the seal.

secant modulus: 14

segregation: A close succession of parallel, relatively narrow, and sharply defined wavy lines of color on the surface of a plastic that differ in shade from surrounding areas and create the impression that the components have separated.

Selectron: Thermoset polyester, PPG Industries, Inc.

   casting compound
   granular molding compound
   sheet molding compound (SMC)

self-extinguishing: A somewhat loosely used term describing the ability of a material to cease burning once the source of flame has been removed.

semiautomatic molding machine: A molding machine in which only part of the operation is controlled by the direct action of a human. The automatic part of the operation is controlled by the machine according to a predetermined program.

semipositive mold: A mold that allows a small amount of excess material to escape when it is closed. (*See also* flash mold; positive mold.)

set: To convert a liquid resin or adhesive into a solid state by curing or by evaporation of solvent or suspending medium or by gelling.

setting temperature: The temperature to which a liquid resin, an adhesive, or products or assemblies involving either is subjected to set the resin or adhesive.

setting time: The period of time during which a molded or extruded product, an assembly, etc., is subjected to heat and/or pressure to set the resin or adhesive.

sharp inside corners: 206

shear rate: The overall velocity over the cross section of a channel with which molten polymer layers are gliding along each other or along the wall in laminar flow:

$$\text{shear rate} = \frac{\text{velocity}}{\text{clearance}} = \frac{\text{cm/s}}{\text{cm}} = \text{s}^{-1}$$

shear strength: 16, 17

shear stress: 16, 17

sheeter lines: Parallel scratches or projecting ridges distributed over a considerable area of a plastic sheet.

shelf life (*see* storage life)

shoe (*see* chase)

Shore hardness: A method of determining the hardness of a plastic material using a scleroscope. This device consists of a small conical hammer fitted with a diamond point and acting in a glass tube. The hammer is made to strike the material under test and the degree of rebound is noted on a graduated scale. Generally, the harder the material, the greater will be the rebound.

short or short shot: In injection molding, failure to fill the mold completely.

shot: The yield from one complete molding cycle, including scrap.

shot capacity: The maximum weight of material that an accumulator can push out with one forward stroke of the ram.

shrink fixture (*see* cooling fixture)

shrink wrapping: A technique of packaging in which the strains in a plastics film are released by raising the temperature of the film thus causing it to shrink over the package. These shrink characteristics are built into the film during its manufacture by stretching it under controlled temperatures to produce orientation of the molecules. Upon cooling, the film retains its stretched condition, but reverts toward its original dimensions when it is heated. Shrink film gives good protection to the products packaged and has excellent clarity.

shrinkage
    choice of values: 5
    from mold dimensions: 4–6
    variation of thermoplastics: 5, 6
    variation of thermosets: 5

Siamese blow: A colloquial term applied to the technique of blowing two or more parts of a product in a single blow and then cutting them apart.

silicone: One of the family of polymeric materials in which the recurring chemical group contains silicon and oxygen atoms as links in the main chain. At present these compounds are derived from silica (sand) and methyl chloride. The various forms obtainable are characterized by their resistance to heat. Silicones are used in the following applications: (a) greases for lubrication; (b) rubberlike sheeting for gaskets, etc.; (c) heat-stable fluids and compounds for waterproofing, insulating, etc.; (d) thermosetting insulating varnishes and resins for both coating and laminating: 185, 186

silk screen printing (screen process decorating): This printing method, in its basic form, involves laying a pattern of an insoluble material, in outline, on a finely woven fabric, so that when ink is drawn across it, it is able to pass through the screen only in the desired areas.

sink mark: A shallow depression or dimple on the surface of an injection molded part due to collapsing of the surface following local internal shrinkage after the gate seals. May also be an incipient short shot.

sinking a mold (*see* hobbing)

sintering: In forming articles from fusible powders, e.g., nylon, the process of holding the pressed-powder article at a temperature just below its melting point for about one-half hour. Particles are fused (sintered) together, but the mass, as a whole, does not melt.

sizing: The process of applying a material to a surface to fill pores and thus reduce the absorption of the adhesive or coating subsequently applied or otherwise to modify the surface. Also, the surface treatment applied to glass fibers used in reinforced plastics. The material used is sometimes called *size*.

sleeve ejector: Bushing-type ejector.

slip additive: A modifier that acts as an internal lubricant that exudes to the surface of the plastic during and immediately after processing. In other words, a nonvisible coating blooms to the surface to provide the necessary lubricity to reduce coefficient of friction and thereby improve slip characteristics.

slip forming: Sheet forming technique in which some of the plastic sheet material is allowed to slip through the mechanically operated clamping rings during a stretch forming operation.

slot extrusion: A method of extruding film sheet in which the molten thermoplastic compound is forced through a straight slot.

slurry preforming: Method of preparing reinforced plastics preforms by wet processing techniques similar to those used in the pulp molding industry.

slush molding: Method for casting thermoplastics in which the resin in liquid form is poured into a hot mold where a viscous skin forms. The excess slush is drained off, the mold is cooled, and the molding is stripped out.

snap-back forming: Sheet forming technique in which an extended heated plastic sheet is allowed to contract over a male form shaped to the desired contours.

solute: The material that dissolves in a solvent.

solution: Homogeneous mixture of two or more components, such as a gas dissolved in a gas or liquid, or a solid in a liquid.

solvation: The process of swelling, gelling, or solution of a resin by solvent or plasticizer.

solvent: Any substance, usually a liquid, which dissolves other substances.

solvent molding: Process for forming thermoplastic articles by dipping a male mold in a solution or dispersion of the resin and drawing off the solvent to leave a layer of plastic film adhering to the mold.

spanishing: A method of depositing ink in the valleys of embossed plastics film.

specific gravity: 6

specific heat: The amount of heat required to raise a specified mass by one unit of a specified temperature.

specific viscosity: The specific viscosity of a polymer is the relative viscosity of a solution of known concentration of the polymer minus one. It is usually determined for a low concentration of the polymer (0.5 g per 100 m of solution or less):

$$n_{sp} = \frac{n - n_0}{n_0} = n_r - 1$$

where n_{sp} = specific viscosity and n_r = relative viscosity.

specific volume, conversion factor: 7

SPI tolerances: 222

spider: (1) In a molding press, that part of an ejector mechanism which operates the ejector pins. (2) In extrusion, a term used to denote the membranes supporting a mandrel within the head/die assembly.

spider lines: Vertical marks on the parison (container) caused by improper welding of several melt flow fronts formed by the legs with which the torpedo is fixed in the extruder head.

spin welding: A process of fusing two objects together by forcing them together while one of the pair is spinning, until frictional heat melts the interface. Spinning is then stopped and pressure held until they are frozen together.

spinneret: A type of extrusion die, i.e., a metal plate with many tiny holes, through which a plastic melt is forced to make fine fibers and filaments. Filaments may be hardened by cooling in air, water, etc., or by chemical action.

spray-up: Covers a number of techniques in which a spray gun is used as the processing tool. In reinforced plastics, for example, fibrous glass and resin can be simultaneously deposited in a mold. In essence, roving is fed through a chopper and ejected into a resin stream, which is directed at the mold by either of two spray systems. In foamed plastics, very fast-reacting urethane foams or epoxy foams are fed in liquid streams to the gun and sprayed on the surface. On contact, the liquid starts to foam.

spreader: A streamlined metal block placed in the path of flow of the plastics material in the heating cylinder of extruders and injection molding machines to spread it into thin layers, thus forcing it into intimate contact with the heating areas.

sprue bushing: A hardened steel insert in an injection mold that contains the tapered sprue hole and has a suitable seat for the nozzle of the injection cylinder. Sometimes called an *adapter.*

sprue gate: A passageway through which molten resin flows from the nozzle to the mold cavity.

sprue lock: In injection molding, the portion of the plastic composition that is held in the cold slug well by an undercut; used to pull the sprue out of the bushing as the mold is opened. the sprue lock itself is pushed out of the mold by an ejector pin. When the undercut occurs on the cavity block retainer plate, this pin is called the *sprue ejector pin.*

spinning: Process of making fibers by forcing plastic melt through spinneret.

spiral flow test: A method for determining the flow properties of a thermoplastic resin in which the resin flows along the path of a spiral cavity. The length of the material that flows into the cavity and its weight give a relative indication of the flow properties of the resin.

spiral mold cooling: A method of cooling injection molds or similar molds wherein the cooling medium flows through a spiral cavity in the body of the mold. In injection molds, the cooling medium is introduced at the center of the spiral, near the sprue section, as more heat is localized in this section.

split cavity: Cavity made in sections.

split-ring mold: A mold in which a split cavity block is assembled in a chase to permit the forming of undercuts in a molded piece. These parts are ejected from the mold and then separated from the piece.

spray coating: Usually accomplished on continuous webs by a set of reciprocating spray nozzles traveling laterally across the web as it moves.

sprayed metal molds: Mold made by spraying molten metal onto a master until a shell of predetermined thickness is achieved. The shell is then removed and backed up with plaster, cement, casting resin, or other suitable material. Used primarily as a mold in sheetforming processes.

stabilizer: An ingredient used in the formulation of some plastics, especially elastomers, to assist in maintaining the physical and chemical properties of the compounded materials at their initial values throughout the processing and service life of the material.

staple: Refers to textile fibers of a short length, usually 0.5−3 in. (12.7−76.2 mm), for natural fibers and sometimes larger for synthetics.

starch: A polysaccharide with the same chemical make-up as cellulose in that it consists of glucose units. It differs from cellulose in manner in which the glucose units are linked together.

stationary platen: The large front plate of an injection molding machine to which the front plate of the mold is secured during operation. This platen does not move during normal operation.

steam molding (expandable polystyrene): Used to mold parts for preexpanded beads of polystyrene using steam as a source of heat to expand the blowing agent in the material. The steam in most cases is contacted intimately with the beads directly or may be used indirectly to heat mold surfaces which are in contact with the beads.

steam plate: Mounting plate for molds, cored for circulation of steam.

stereospecific plastics: Implies a specific or definite order of arrangement of molecules in space. This ordered regularity of the molecules in contrast to the branched or random arrangement found in other plastics permits close packing of the molecules and leads to high crystallinity, i.e., as in polypropylene.

stir-in resin: A vinyl resin that does not require grinding to affect dispersion in a plastisol or organisol.

stitching: The progressive welding of thermoplastic materials by successive applications of two small mechanically operated electrodes, connected to the output terminals of a radiofrequency generator, using a mechanism similar to that of a normal sewing machine.

storage life: The period of time during which a resin or packaged adhesive can be stored under specified temperature conditions and remain suitable for use. Storage life is sometimes called *shelf life.*

strain: 14

strength, equivalent of thick parts: 208, 209

stress

    allowable working: 14, 227, 228, 229

    concentration factor: 206, 237

    crack, environmental: 42

    intermittent load: 229

    relaxation: 37

    −strain curve, comparison of steel and plastics: 10

    weld line: 215

stretch forming: A plastic sheet forming technique in which the heated thermoplastic sheet is stretched over a mold and subsequently cooled.

striation: Rippling of thick parisons caused by local orientation effect in the melt by the spider legs.

stripper-plate: A plate that strips a molded piece from core pins or force plugs. The stripper-plate is set into operation by the opening of the mold.

strongly negative elements: Are called nonmetallic. They have a tendency to acquire electrons.

strongly positive elements: Are termed metallic. They have a tendency to give up electrons.

Styrafil: Reinforced polystyrene, GP, Fiberfil Division, Dart Industries.

styrene: 178, 179, 180

styrene, acrylonitrile: 187, 188, 189

styrene−rubber plastics: Plastics consisting of at least 50% of a styrene plastic combined with rubbers and other compounding ingredients.

Styron: Polystyrene, GP, Dow Chemical Co. polystyrene, impact

Styropor: Polystyrene, expandable bead, BASF, Wyandotte Corp.

submarine gate: A type of edge gate where the opening from the runner into the mold is located below the parting line or mold surface as opposed to conventional edge gating where the opening is machined into the surface of the mold. With submarine gates, the item is broken from the runner system on ejection from the mold.

Sufil: Reinforced polysulfone, Fiberfil Division, Dart Industries.

sunlight simulated test: 40, 41

Super Dylan: Polyethylene, ARCO/Polymers. Type II, type III, type IV.

Superflex: Polystyrene, impact, Hammond Plastics Division

Superflow: Polystyrene, GP, Hammond Plastics Division

supersaturation: A solution containing more of the solute than would normally dissolve.

support post or pillar: Posts used in molds to resist deflection under pressure.

surface resistivity: The electrical resistance between opposite edges of a unit square of insulating material. It is commonly expressed in ohms. (Also covered in ASTM D257-54T.)

surface treating: Any method of treating a polyolefin so as to alter the surface and render it receptive to inks, paints, lacquers, and adhesives such as chemical, flame, and electronic treating.

surging: Unstable pressure build-up in an extruder leading to variable throughput and waviness of the parison.

Surlyn: Ionomer, Du Pont Co.: 108, 109, 110

suspension: A mixture of fine particles of any solid with a liquid or gas. The particles are called the *disperse phase,* the suspending medium is called the *continuous phase.*

syndiotactic: A chain of molecules in which the methyl groups alternate regularly on opposite sides of the chain.

synergism: A term used to describe the use of two or more stabilizers in an organic material where the combination of such stabilizers improves the stability to a greater extent than could be expected from the additive effect of each stabilizer.

tab gated: A small removable tab of approximately the same thickness as the mold item, usually located perpendicular to the item. The tab is used as a site for edge gate location, usually on items with large flat areas.

tack range: The period of time in which an adhesive will remain in the tacky-dry condition after application to an adherend, under specified conditions of temperature and humidity.

taper: 214

Technyl: Nylon type 66, Rhone-Poulenc Polymers.

teeth
number in gear: 235, 236
thickness at pitch line: 236, 239, 240

Teflon: Fluoropolymers, Du Pont Co.
FEP
TFE
PFA

Tefzel: ETFE fluoropolymer, Du Pont Co.: 106

tenacity (gpd): The term generally used in yarn manufacture and textile engineering to denote the strength of a yarn or a filament for its given size. Numerically it is the grams of breaking force per denier unit of yarn or filament size; grams per denier, gpd. The yarn is usually pulled at the rate of 12 in./min. Tenacity equals breaking strength (g) divided by denier. (Tenacity, gpd) (Specific Gravity) (12,800) = (Tensile Strength, psi).

Tenite: Celluosics, Eastman Chemical Products Co.: 93, 94, 95
polyallomer
thermoplastic polyester, PBT
polyethylene type I
polypropylene

Tenite, polypropylene: 172

tensile bar (specimen): 11

tensile impact strength: 18, 19

tensile strength: 9
influence of rate of jaws, temperature: 11
usefulness of data: 12

terephthalate: A compound used in manufacture of linear crystalline polyester resins, fibers, and films by combination with glycols.

terpolymer: 52

Texin: Thermoplastic urethane, Mobay Chemical Corp.

Therimage: A trademark for a decorating process for plastic which transfers the image of a label or decoration to the object under the influence of heat and light pressure.

thermal conductivity: Ability of a material to conduct heat; physical constant for quantity of heat that passes through unit cube of a substance in unit of time when difference in temperature of two faces is 1°.

thermal expansion (coefficient of): 24

thermal stress cracking (TSC): Crazing and cracking of some thermoplastic resins which results from overexposure to elevated temperatures.

thermoforming: 256, 257

thermoforms: The product that results from a thermoforming operation.

thermoplastic: *(adj)* Capable of being repeatedly softened by heat and hardened by cooling. *(n)* A material that will repeatedly soften when heated and harden when cooled. Typical of the thermoplastics family are the styrene polymers and copolymers, acrylics, cellulosics, polyethylenes, vinyls, nylons, and the various fluorocarbon materials.

thermoplastic copolyester: 151, 152

thermoplastic elastomers: 197, 198

thermoplastic polymers: 57, 58

thermoplastic urethane: 182

thermoplastics vs thermosets, comparisons: 199, 200, 201

thermoset: A material that will undergo or has undergone a chemical reaction by the action of heat, catalysts, ultraviolet light, etc., leading to a relatively infusible state. Typical of the plastics in the thermosetting family are the aminos (melamine and urea), most polyesters, alkyds, epoxies, and phenolics.

thermosets vs thermoplastics: 199, 200
comparisons: 201

thermosetting polymers: 56

thickness
tolerance on deep walls: 216
variation: 209
variation of deep walls: 216
wall or part: 208, 209, 211

thick wall, conversion to favorable dimensions: 211, 212

thixotropic: Said of materials that are gellike at rest, but fluid when agitated. Liquids containing suspended solids are apt to be thixotropic. Thixotropy is desirable in paints.

thread contour (internal): 218

threading in plastics: 218, 219, 220

three plate mold: A third or intermediate movable plate used in injection molds to make possible center or offset gating of each cavity.

tie bars: Bars that provide structural rigidity to the clamping mechanism often used to guide platen movement.

toggle action: A mechanism that exerts pressure developed by the application of force on a knee joint. It is used as a method of closing presses and also serves to apply pressure at the same time.

tolerances, manner of specifying: 222, 223

Torlon: Poly(amide-imide), Amoco Chemicals Corp.: 136, 137 torpedo (or spreader): A streamlined metal block placed in the path of flow of the plastics material in the heating cylinder of extruders and injection molding machines to spread it into thin layers, thus forcing it into intimate contact with heating areas.

torsion: Stress caused by twisting a material.

TPX: 165

tracking: A phenomenon wherein a high-voltage-source current creates a leakage or fault path across the surface of an insulating material by slowly but steadily forming a carbonized path.

transfer molding: A method of molding thermosetting materials, in which the plastic is first softened by heat and pressure in a transfer chamber, then forced through high pressure through suitable sprues, runners, and gates into closed mold for final curing.

translucent: Description of a material or substance capable of transmitting some light, but not clear enough to be seen through.

transmittance, luminous of transparent plastics: 8

transparency, plastics: 2

transparent: Descriptive of a material or substance capable of a high degree of light transmission (e.g., glass). Some polypropylene films and acrylic moldings are outstanding in this respect.

trioxane: Used in organic synthesis; disinfectant; nonluminous odorless fuel.

Trogamid: Nylon type 63T, Kay-Fries, Chemicals.

tumbling: Finishing operation for small plastic articles by which gates, flash, and fins are removed and/or surfaces are polished by rotating them in a barrel together with wooden pegs, sawdust and polishing compounds.

two-level mold: Also called stack-mold. Placement of one cavity of a mold above another instead of alongside to reduce clamping force.

Tyril: SAN copolymer, Dow Chemical Co.

Udel: Polysulfone, Union Carbide Corp.: 190–192

Ultem: 154–157

Ultra-high-molecular-weight polyethylene: 161, 162

Ultrasonic sealing and welding: 251, 252, 253

Ultrathene: EVA copolymers, U. S. Industrial Chemicals Co.

ultraviolet: Zone of invisible radiations beyond the violet end of the spectrum of visible radiations. Since UV wavelengths are shorter than the visible, their photons have more energy, enough to initiate some chemical reactions and to degrade most plastics.

undercut: 220, 221

gear teeth: 235

molding: 220

unnotched impact strength: 206 (*see also* data sheet of each material)

unsaturated compounds: 50

unsaturation: Implies a condition of partially filled. In terms of a compound, it denotes a condition of being able to take up or react with more atoms.

urea, formaldehyde resin (urea resin): 89

urethane: 181, 182

UV stabilizer (ultraviolet): Any chemical compound that, when admixed with a thermoplastic resin, selectively absorbs UV rays.

vacuum forming: Method of sheet forming in which the plastic sheet is clamped in a stationary frame, heated, and drawn down by a vacuum into a mold. In a loose sense, it is sometimes used to refer to all sheet forming techniques, including *drape forming* involving the use of vacuum and stationary molds.

vacuum metallizing: Process in which surfaces are thinly coated with metal by exposing them to the vapor of metal that has been evaporated under vacuum (one millionth of normal atmospheric pressure).

valence: 48

valence, ionic: 48

Valox: Thermoplastic polyester, PBT, General Electric Co.: 147

Vespel: Polyimide, Du Pont

polyamide, aromatic

vinyl resin: A synthetic resin formed by the polymerization of chemical compounds containing the group $CH_2 = CH—$. In particular, polyvinyl chloride, acetate, alcohol and butyral are referred to.

viscoelasticity: 31

viscosity: Internal friction or resistance to flow of a liquid. The constant ratio of shearing stress to rate of shear. In liquids for which this ratio is a function of stress, the term " apparent viscosity" is defined as this ratio.

volume resistivity (specific insulation resistance): The electrical resistance between opposite faces of a 1-cm cube of insulating material. It is measured under prescribed conditions using a direct current potential after a specified time of electrification. It is commonly expressed in ohm-centimeters. The recommended test is ASTM D257-54T.

vulcanization: The chemical reaction that induces extensive changes in the physical properties of a rubber and that is brought about by reacting the rubber with sulfur and/or other suitable agents. The changes in physical properties include decreased plastic flow, reduced surface tackiness, increased elasticity, much greater tensile strength, and considerably less solubility. More recently, certain thermoplastics, e.g., polyethylene, have been formulated to be vulcanizable. Cross-linking is encouraged, thereby giving resistance to deformation of flow above the melting point.

vulcanized fiber: 196

Vydyne: Nylon, Monsanto Co.: 114
   type 66
   type 69
   copolymer

wall thickness: 208, 209, 210, 211, 212, 213, 216

warpage: Dimensional distortion in a plastic object after molding.

water absorption, plastics: 7

weathering, accelerated test: 40

weathering, outdoor test: 41

weatherometer: An instrument that is utilized to subject articles to accelerated weathering conditions, e.g., rich UV source and water spray.

weight loss on heating: 41, 42

weight, plastics, consideration: 1

weld lines: 215

weld lines, location for strength: 215

welding by spinning: 215

weld mark (also flow line): 215

Wellamid: Nylon, Wellman, Inc.
   type 66
   type 6

wet strength: The strength of paper when saturated with water, especially used in discussions of processes whereby the strength of paper is increased by the addition, in manufacture, of plastics resins. Also, the strength of an adhesive joint determined immediately after removal from a liquid in which it has been immersed under specified conditions of time, temperature, and pressure.

window: A defect in a thermoplastics film sheet or molding caused by the incomplete "plasticization" of a piece of material during processing. It appears as a globule in an otherwise blended mass. (*See also* fish eye.)

working life (stress): 227

working performance duration: 227

Xylon: Reinforced nylon type 66, Fiberfil Division, Dart Industries.

Yield, flexural strength: 15

Yield value (yield strength): 12, 13

Young's modulus of elasticity: 12, 13

Zytel: Nylon, Du Pont Co., 113
   copolymer
   type 66
   type 612